P.

The Nature of Nature

"Vandana's is a clear, indefatigable voice of outstanding intellect and compassion, of deliberate and compelling outrage in our planetary crisis. She is a privilege to know and learn from in my lifetime."

—Professor Marilyn Waring, author of *If Women Counted: A New Feminist Economics*

"Vandana Shiva's pioneering efforts to expose how a GMO- and fossil fuel–based petrochemical agricultural system has wreaked havoc on our species' global food chain and undermined ecosystems around the world have touched off a worldwide conversation. She has singlehandedly drawn several generations into regenerative agriculture and ecosystem restoration, particularly in developing countries. Her new book makes the incontrovertible connection between a warming climate and an outmoded agricultural system and guides our species into a more ecologically sensitive approach to provisioning food by treating nature as a shared commons for all of life on Earth."

—Jeremy Rifkin, economic and social theorist; writer; activist

ALSO BY VANDANA SHIVA

Terra Viva: My Life in a Biodiversity of Movements (2022)
Oneness vs. the 1%: Shattering Illusions, Seeding Freedom (2018)
Who Really Feeds the World? (2016)
Seed Sovereignty, Food Security: Women in the Vanguard (ed.) (2015)
Making Peace with the Earth: Beyond Resource, Land and Food Wars (2012)
Soil Not Oil: Climate Change, Peak Oil and Food Insecurity (2009)
Globalization's New Wars: Seed, Water & Life Forms (2005)
Water Wars: Privatization, Pollution and Profit (2002)
Stolen Harvest: The Hijacking of the Global Food Supply (2000)
Biopiracy: The Plunder of Nature and Knowledge (1997)
Ecofeminism (co-authored with Maria Mies, 1993)
Monocultures of the Mind: Biodiversity, Biotechnology and Agriculture (1993)
The Violence of the Green Revolution: Ecological Degradation and Political Conflict in Punjab (1992)
Staying Alive: Women, Ecology and Survival in India (1988, 2010)

The Nature of Nature

THE METABOLIC DISORDER OF CLIMATE CHANGE

Vandana Shiva

Chelsea Green Publishing
White River Junction, Vermont, USA
London, UK

Copyright © 2024 by Vandana Shiva.
All rights reserved

No part of this book may be transmitted or reproduced in any form by any means without permission in writing from the publisher.

The Nature of Nature: The Metabolic Disorder of Climate Change was originally published in India in 2024 by Women Unlimited Ink, 7/10, First Floor, Sarvapriya Vihar New Delhi – 110 016.

This edition published by Chelsea Green Publishing, 2024.

Typeset by Manmohan Kumar, Delhi – 110 035.

Printed in Canada.
First printing October 2024.
10 9 8 7 6 5 4 3 2 1   24 25 26 27 28

Our Commitment to Green Publishing
Chelsea Green sees publishing as a tool for cultural change and ecological stewardship. We strive to align our book manufacturing practices with our editorial mission and to reduce the impact of our business enterprise in the environment. We print our books using vegetable-based inks whenever possible. This book may cost slightly more because it was printed on paper that contains recycled fiber, and we hope you'll agree that it's worth it. *The Nature of Nature* was printed on paper supplied by Marquis that is made of recycled materials and other controlled sources.

ISBN: 978-1-64502-287-9 (paperback)
ISBN: 978-1-64502-288-6 (ebook)
ISBN: 978-1-64502-289-3 (audiobook)

Library of Congress Control Number: 2024944663

Chelsea Green Publishing
White River Junction, Vermont, USA
London, UK
www.chelseagreen.com

Contents

1. A Planetary Crisis: the Biodiversity, Climate, Food Connection   1
2. Two Paradigms: Mechanophilia vs. Nature's Technology   19
3. Industrial Farming and the Illusion of Food Security   39
4. Dead Food, Dead Metabolism   63
5. The Fake Food Dystopia   95
6. The Future of Food   135

Acknowledgements   161

1 | A Planetary Crisis
the Biodiversity, Climate, Food Connection

The extinction emergency, climate havoc and climate chaos, and the food crisis are the symptoms and consequences of violence and war against the earth and earth citizens unleashed by the greed of the 1%. This 1% extracts, encloses and pollutes a sentient environment, destroying the conditions of life on earth by appropriating the resources that sustain people's livelihoods. The 'fossilised' construct of a 'dead' earth, combined with the economy of extraction and enclosures, has created the multidimensional emergency that threatens our future.

Our age is frequently referred to as the Anthropocene. I do not accept this term, because all of humanity is not predatory. Humans as a species have not caused climate disasters or the extinction crisis—it is the exploitative, unchecked practice of the 1% that has done so. These crises are not the anthropogenic impact of actions caused by all of humanity; they are the capitalogenic impact of the reckless actions of the 1%. I also do not use the term *Anthropocene* because we need to move beyond anthropocentrism if we want to cultivate a future with all life on earth. The earth is for all beings, not just for human beings.

The top 1% of emitters produces over 1,000 times more pollution than the bottom 1%. As a 2023 Oxfam report indicates, carbon emissions by the 1% are greater than the emissions of the poorest two-thirds of humanity.[1] In the absence of both

---

[1] 'Richest 1% Emit as Much Planet-Heating Pollution as Two-Thirds of Humanity', Oxfam International, November 20, 2023, https://www.oxfam.org/en/press-releases/richest-1-emit-much-planet-heating-pollution-two-thirds-humanity.

experience and knowledge of the ecological and social impact of the 'greed economy', as well as the lack of differentiation between real solutions to real ecological problems, democratic rejection of the rule of the 1% is morphing into a denial of the serious ecological crises that threaten the lives of diverse species and vulnerable members of the human community.

Climate disasters add to the destructive impact of colonialism and maldevelopment which place profits above nature and people. A new green colonialism is emerging through greenwash—reducing a complex, interrelated ecological crisis to distinct and disconnected crises and one-dimensional symptoms, then blindly promoting false solutions for more profit and greater control over the earth, its resources and our lives.

It is the countries in the South that disproportionately pay the highest price for ecological destruction while having contributed the least to it, experiencing the worst impact of floods and droughts, cyclones and heat waves. I have worked with communities affected by the Odisha Super Cyclone in 1999, in which 10,000 people lost their lives; the Kedarnath disaster in 2013, in which 6,054 people died and the Rishi Ganga disaster in 2021, in which 250 people died.

Environmental crises invite us to go beyond the anthropocentric arrogance that drives the war against the earth and makes the 1% indifferent to the destruction of diversity and ecological processes. But, to make matters worse, polluters are expanding and accelerating the destruction by taking over international environmental treaties that were actually created to regulate their practices. Instead, they are mutating the treaties into instruments for creating new markets in pollution and environmental damage.

Three decades of the international climate treaty

Ecological movements have grown since the 1970s in response to ecological destruction driven by an extractive model of the economy as 'development', as 'growth', as corporate globalisation. The destruction of biodiversity in forests, farms and oceans over the last four decades through industrial, monoculture forestry, farming and fisheries led to the emergence of movements to protect it. The pollution of the air and atmosphere, culminating in a destabilised climate, climate extremes and climate havoc as a result of pollution from fossil fuels and toxic chemicals derived from them, saw the introduction of two international environment treaties signed at the Earth Summit in Rio in 1992 by governments of the world: these were the Convention on Biological Diversity (CBD) to conserve and protect biodiversity and the UN Framework Convention on Climate Change (UNFCCC). Both treaties are interconnected because the biosphere and the atmosphere are interconnected.

The Earth Summit happened before the meeting in Marrakesh in 1994, where the World Trade Organization (WTO) was set up. It was held, in a pre-globalisation, pre-corporate control era, around pressing ecological concerns, where eco-movements compelled national governments as well as international agencies to commit to the protection of the environment and of indigenous people.

The UN system created after WWII was based on 'one country, one vote'. At the Earth Summit, both the biodiversity as well as climate agreements were shaped by the countries of the South, because they are not only home to a majority of the 36 biodiversity hotspots essential to survival but also home to over two billion people, including some of the world's poorest, who rely directly on healthy ecosystems for their livelihoods and well-being.

The CBD was intended to protect and conserve biodiversity, the knowledge of indigenous people and the sovereignty of countries. Over time, this Convention has been completely undermined as regulations to prevent biopiracy are being subverted, biosafety regulations are being bypassed through digital mapping and gene-edited GMOs and biodiversity destruction is being concealed under 'biodiversity offsets'. The mutation of the international environmental treaties, meant to address the planetary ecological crisis, is thus taking place at both the ecological and political-economic levels. Today, the international is no longer intergovernmental; it has become the space controlled by globalists—the 1%.

Over 30 years have elapsed since the wake-up call at the Earth Summit, and biodiversity erosion has now become an extinction emergency. The climate crisis has become a climate catastrophe, despite the annual Climate Change Conference of the Conference of the Parties (COP) that meets to discuss strategies and progress in coping with climate change.

The slippery slope of deregulation and corporatisation

Climate change is an issue of justice, and of life and death. The objective of the UN climate treaty was to stop pollution and climate injustice, and it was *legally binding*. Polluters must stop polluting. Polluters must pay. Since the industrialised countries were responsible for the pollution caused by fossil fuels, the emission reduction targets of the treaty originally applied to 37 industrialised nations, identified as Annex B countries at COP 3 in Kyoto in 1997. The first phase of the Kyoto Protocol (adopted in 1997 but enforced since 2005) required rich nations, the historic polluters, to reduce emissions by 5 percent, compared to 1990 levels, between 2008 and 2012. However, the polluters transformed these legally binding restrictions on pollution and emissions into trade in pollution through the Doha Amendment of the Kyoto Protocol in 2012.

The two most significant COP meetings were held in Copenhagen and Paris in 2009 and 2015, respectively. In 2009, US President Barack Obama flew to Copenhagen, proposed a dismantling of the legal framework and its substitution of voluntary commitments with a small group of countries outside the conference negotiations, held a press conference and then flew away. That is why President Evo Morales of Bolivia stood up in the negotiating hall and said, "We are here to protect the Rights of Mother Earth, not the Rights of Polluters." He took the initiative to mobilise citizens of the world to draft a declaration on the Rights of Mother Earth, a process I was a part of.[2]

I wrote *Soil Not Oil* for the Copenhagen Summit. At that time we also conducted participatory research, entitled *Climate Change at the Third Pole*, and a pilgrimage across the Western Himalayas to assess the impact of climate change on Himalayan communities and ecosystems. The people of the Himalaya have not contributed to the pollution that is melting their glaciers and threatening their lives through disasters.[3]

---

[2] For more, read Vandana Shiva, *Reclaiming the Commons: Biodiversity, Indigenous Knowledge, and the Rights of Mother Earth* (Santa Fe: Synergetic Press, 2020); Vandana Shiva, *Origin: The Corporate War on Nature and Culture* (Delhi: Natraj Books, 2018).

[3] Vandana Shiva and Vinod Kumar Bhatt, *Climate Change at the Third Pole: The Impact of Climate Instability on Himalayan Ecosystems and Himalayan Communities* (New Delhi: Navdanya/RFSTE, 2009).

A Planetary Crisis

COP 21, held in Paris in 2015, marked the end of a legally binding framework. Paris was all about 'voluntary' commitments. More significantly, it marked the end of UN agreements as agreements between countries, through their elected governments, accountable to the people. The Paris Agreement shifted the goal from the concrete and legally binding objective of polluters reducing emissions to 196 countries deciding to make voluntary promises to keep the rise in average global temperatures well below 2°C. Paris also began a new process of 'outcomes' and 'decisions' led by billionaires like Bill Gates, outside and distinct from the formal negotiations between governments.

The 2023 COP 28 in Dubai was presided over by Sultan Ahmed Al Jaber, head of the Abu Dhabi National Oil Company (ADNOC), in which BlackRock Inc., Eni SpA and KKR & Co. Inc. have global investments. Al Jaber is also the chairman of Masdar, the leading building and construction materials company in Saudi Arabia. It was the first time in the history of the Climate Convention—whose aim is to reduce emissions from fossil fuels—that the CEO of an oil giant presided over negotiations. Ironically, the meeting for reducing emissions was organised in the oil capital of the world, and the future of agriculture was deliberated over in the desert.

Corporations that are driving fossil fuel pollution, both through direct use and through industrialised, chemical intensive agriculture, were the dominating presence in Dubai. Although food and agriculture have so far been ignored at most COP conferences, the corporate capture of the food and agriculture agenda was plainly evident at COP 28.

The Climate Convention opened with a special session dedicated to COP 28 UAE Declaration on Sustainable Agriculture,

Resilient Food Systems, and Climate Action. Leaders of 160 countries endorsed the landmark declaration to help strengthen food systems, build resilience to climate change, reduce global greenhouse gas emissions and contribute to the global fight against hunger.[4] The UAE pledged only US$100 million, while at the same time committing US$30 billion to a new Dubai-based Green Investment private equity fund, Alterra, that will partner with BlackRock and other asset management firms to make 'climate investments' in the South.[5]

Agribusiness, represented by ADM, Bayer, Cargill, Danone, Nestlé, Olam Agri, Syngenta and Google, which have systematically destroyed biodiversity in the soil and the environment, launched an initiative to mislead people into thinking that it will contribute to what it calls 'regenerative agriculture'. The Dairy Methane Action Alliance was formed by Big Dairy—Danone, General Mills, Kraft Heinz, Lactalis USA, Bel Group and Nestlé; Big Ag; and Big Food—Bayer, Cargill, Danone, Louis Dreyfus, Nestlé, Olam, PepsiCo, Tyson and Yara—announcing an initiative to 'decarbonise' the industrial food chain, even as their operations contribute 50 percent of the pollution associated with industrial food systems.[6]

As is the practice now, at the end of the session on agriculture, Bill Gates walked onto the stage to announce a partnership

---

[4] COP28 UAE Declaration on Sustainable Agriculture, Resilient Food Systems, and Climate Action, https://www.cop28.com/en/food-and-agriculture.

[5] Joe Lo, 'What Is Alterra, the UAE's $30 Billion Green Investment Fund?', Climate Home News, December 10, 2023, https://www.climatechangenews.com/2023/12/10/what-is-alterra-the-uaes-30-billion-green-investment-fund.

[6] 'The Davos-isation of the Climate COP', GRAIN, February 15, 2024, https://grain.org/en/article/7104-the-davos-isation-of-the-climate-cop.

between the UAE and the Bill & Melinda Gates Foundation with a US$200 million fund for food systems, agriculture innovations and climate action. This fund will focus on agricultural research, scaling agricultural innovations and funding technical assistance for implementing the COP declaration. As *The Guardian* (also funded by Gates) gushed, 'Food is Finally on the Table'.[7]

This book looks into the root causes of climate change, explores the intimate links between our food and the climate and enquires whether Bill Gates's fake food 'innovation' can be a solution to global malnutrition, hunger and climate change, or whether it will exacerbate the crises. It also offers alternatives that work with nature, according to nature's ecological laws, and are, in fact, the real solutions to climate change that also regenerate the earth and address food security.

Big money hijacks the climate agenda

Bill Gates is not government—he is not a 'party' in UN negotiations—but in Paris, for the first time, he emerged as the 'master' of global platforms, using the COP to promote geo-engineering, genetic engineering and now fake food, Net Zero and carbon offsets. Through manipulation, Gates has replaced elected governments and displaced democracy. He has substituted the principle of 'polluter pays' with 'polluter gets paid', using false climate solutions to create new markets, enclose new commons and find new ways to make money.

---

[7] Whitney Bauck, '"Food Is Finally on the Table": Cop28 Addressed Agriculture in a Real Way', *The Guardian*, December 17, 2023, https://www.theguardian.com/environment/2023/dec/17/cop28-sustainable-agriculture-food-greenhouse-gases?ref=upstract.com.

In the years following Paris, the agenda of false solutions has gained firm ground. Gates is blaming nature and farmers, pushing anti-nature, anti-farmer remedies which will only deepen the social and ecological crises. The sun is not the problem, pollution is; so 'dimming the sun' by injecting aerosols in the atmosphere is not going to solve the issue of global warming. Cows are not Enemy No. 1, factory farms are. Feeding cows an energy intensive feed of corn and soyabean is a major source of pollution, not the methane that is produced when cows digest the feed. Trees are not the problem, so 'fake trees' can't be a solution to climate change. At the Climate Forward event in September 2023 in New York, when David Gelles of the *New York Times* asked Bill Gates about trees sequestering carbon dioxide, he declared that those who think trees absorb carbon dioxide are idiots: 'Are we the science people or are we the idiots?'[8] Gates promotes industrial, mechanical systems of carbon capture as a technological innovation. But fake trees, which capture carbon mechanically, cannot perform photosynthesis; they cannot produce food and fiber; they cannot give or sustain life—they cannot create humus and living soil; they cannot regenerate and conserve water. It is a mechanical mind that sees *one* function only and tries to substitute it without taking into consideration a diversity of functions. Net Zero is not zero emissions; it is about making more money through clever financial jugglery.

---

[8] 'Can Planting Trees Really Help Fight Climate Change?', Al Jazeera, October 3, 2023, https://www.aljazeera.com/news/2023/10/3/can-planting-trees-really-help-fight-climate-change; Gabriel Labbate, 'Bill Gates Made Waves with His Statements on Climate Change. Here's Why He's Right—and What Most People Missed', *Fortune*, November 16, 2023, https://fortune.com/2023/11/16/bill-gates-climate-change-trees-environment-un-gabriel-labbate/?.

A Planetary Crisis

Gates has admitted that getting to zero doesn't really mean zero:

> There are no realistic paths to zero that involve abandoning these fuels completely or stopping all the other activities that also produce greenhouse gases (like making cement, using fertilizer, or letting methane leak out of natural gas power plants). Instead, in all likelihood, in a zero-carbon future we will still be producing some emissions, but we'll have ways to remove the carbon they emit.[9]

The Net Zero Initiative, chaired by former Bank of England governor Mark Carney, talks of 'changing the plumbing of the whole financial system forever', but really it's just another way by which, riding on fake economics and fake science, the 1% can make more money while 'offsetting' pollution through 'carbon credits'. The finance and accounting firms are ready with the financial infrastructure for Net Zero. According to a McKinsey report on Net Zero transition: 'We estimate that the cumulative capital spending on physical assets for the net-zero transition between 2021 and 2050 would be about $275 trillion.'[10]

Biodiversity, climate, food, health

The biodiversity crisis, the climate crisis and the food and health crises are a single planetary crisis, because the biosphere and

---

[9] Bill Gates, *How to Avoid a Climate Disaster: The Solutions We Have and the Breakthroughs We Need* (New York: Alfred A. Knopf, 2021), 19.

[10] McKinsey Global Institute, *The Net-Zero Transition: What It Would Cost, What It Could Bring*, January 2022, https://www.mckinsey.com/capabilities/sustainability/our-insights/the-net-zero-transition-what-it-would-cost-what-it-could-bring.

atmosphere are an intimately coupled system of the living earth. The biosphere has created and regulates the earth's climate system. The biosphere, in turn, is sustained through food cycles and the flow of food as the currency of life across species and ecosystems. The carbon cycle is a food cycle. What flows across living systems is nutrition. The nutrient cycle is the foundational cycle of life. It begins with the absorption of carbon dioxide from the atmosphere, with the help of sunlight through photosynthesis. Atmospheric carbon is transformed into carbohydrates by plants. The carbon thus returns to the biosphere, including the biodiversity of plants and the biodiversity in the soil. Animals, including humans, eat the plants as food and emit carbon dioxide. This is the carbon cycle. Climate change is a result of the rupture of this cycle, caused by fossil fuels.

The shift from biodiversity-based food systems to oil-based, fossil fuel- and fossil chemical-based food systems has violated the earth's ecological cycles, created a paradigm of linear extractivism and the associated creation of waste, contributing to the pollution of water, of the soil, of the atmosphere and of our food. The earth's capacity to regulate her climate through the biosphere and biodiversity is being disrupted by pollution from the burning of fossil fuels and using their products in the form of petrochemicals. This pollution creates what are referred to as greenhouse gases (GHGs) that have been increasing since the industrial age.

The pollution of the atmosphere through GHG emissions—CO_2 (carbon dioxide), N_2O (nitrous oxide), CH_4 (methane)—is what drives climate change, and industrialised, globalised food production is responsible for 50 percent of GHG emissions. The Poison Cartel has already trapped farmers worldwide in an energy intensive, chemical intensive, capital intensive agricultural system that is causing a deep agrarian crisis, food crisis and health crisis. The Age of Oil has totally transformed our food systems. We are actually eating oil, from the production of food to its industrial processing, plastic packaging and distribution. The junk energy of fossil fuels has not only negatively impacted the earth's metabolism and led to climate havoc, but junk and ultra-processed foods have disrupted human metabolism and led to a pandemic of chronic diseases as well.

Climate havoc has made natural disasters like floods and droughts more common, and more extreme, leading to frequent crop failures and intense food insecurity, because unlike indigenous, diverse, artisanal farming, industrial monocultures are more vulnerable to destruction. Global estimates reveal that by 2050, 3.5 billion people will suffer from food insecurity, an increase of 1.5 billion people over today.[11] Rise in temperature, combined with the destabilisation of the hydrologic cycle, has adversely impacted our food systems. Between 2021–22, farmers in the Bay of Bengal were hit by multiple cyclones—Yaas, Gulab, Jawad, Asani and Sitrang—that completely destroyed their standing crops. In 2023, there was no rain, which led to a drought, affecting the sowing of paddy and tubers like turmeric and colocasia. Several districts in the desert state of Rajasthan were battered by the tropical cyclone Biparjoy on June 18, 2023, significantly wiping out local bird and animal populations, which in turn caused severe damage to crops, as

---

[11] 'IEP: Over One Billion People at Threat of Being Displaced by 2050 Due to Environmental Change, Conflict and Civil Unrest', PRNewswire.com, September 9, 2020, https://www.prnewswire.com/in/news-releases/iep-over-one-billion-people-at-threat-of-being-displaced-by-2050-due-to-environmental-change-conflict-and-civil-unrest-829796933.html.

harmful insect populations grew unchecked. In the Doon Valley, in the hill state of Uttarakhand, where a decade ago there was a flourishing growth of pulses, such as urad, navrangi, masoor and moong, today they are nearly extinct because of extreme rain. In 2024, the failure of the winter rains here has already ruined the rabi crop of wheat and mustard. The Vidarbha region of Maharashtra, too, showed signs of climate change in 2023, as erratic monsoon rains dumped half the year's precipitation in *one day*. Heavy rains destroyed soyabean and cotton, as 35 percent of the area could not even begin the sowing process.

Food is at the heart of the climate debate at present both because of the impact of climate catastrophes on agriculture and due to the concerted efforts of the 1% to eradicate small farms and farmers by aggressively funding fossil fuel and tech-propelled food production. Bill Gates and the tech giants of Silicon Valley are heavily investing in fake food businesses and also buying farmland. In fact, Gates is now the biggest landowner in the US.[12]

The false solution to climate change, being promoted in the form of fake food made in labs, is creating a dystopia of farming without farmers, and food without farms. However, lab food requires more resources for feedstock and, being resource and energy intensive, *contributes to higher GHG emissions*. Rushing faster and further down the path of resource and energy

---

[12] Darren Orf, 'The Truth about Why Bill Gates Keeps Buying Up So Much Farmland', *Popular Mechanics*, January 18, 2023, https://www.popularmechanics.com/science/environment/a42543527/why-is-bill-gates-buying-so-much-farmland; see also Nick Estes, 'Bill Gates Is the Biggest Private Owner of Farmland in the United States. Why?', *The Guardian*, April 5, 2021, https://www.theguardian.com/commentisfree/2021/apr/05/bill-gates-climate-crisis-farmland.

A Planetary Crisis

intensive industrial food production, processing and distribution will only increase centralisation and corporate control of the food system and accelerate the destabilisation of the earth and climate systems.

There is an alternative path, a path made by walking with the earth, following the ecological laws of the earth—the law of diversity and the law of return, shortening the distance between producers and consumers, deindustrialising and deglobalising food systems to reduce emissions and enhance health. This path offers solutions to the climate crisis, the extinction crisis and the hunger and health crises, because the health of the planet and our health are interconnected.

Regenerating the earth through care is our ethical, ecological duty. In regeneration lies the potential, power and promise of healing the earth and humanity. Ecological laws have sustained life on earth through its diverse evolutionary stages, increasing recycling through circular economies based on biodiverse, chemical-free, artisanal, local food systems. The same processes, based on regenerating biodiversity, that produce healthy food also address climate change by getting rid of emissions from fossil fuels and fossil chemicals used in energy and chemical intensive production, long-distance transport and industrial processing. The ecological, democratic and humane option for addressing climate change is growing, as is eating real, healthy food by creating biodiversity and ecological, local, circular, living food economies. The artificial solutions offered by the food industry will increase hunger by diverting food from people to feedstock for producing lab food, just as the diversion of food to animal feed and biofuel has exacerbated it. It will aggravate climate change through increased energy use and increase disease through the ultra-ultra-processing of food using synthetic ingredients.

Fossil fuel/chemical-free farming and returning organic matter to the soil allow soil biodiversity to flourish, as the symbiosis between plants and soil organisms, like mycorrhizal fungi, produces healthier food. When fungi feed the plants with minerals, photosynthesis increases, allowing more food to grow while also feeding the soil organisms. Degenerative cycles are transformed into regenerative cycles. The web of life feeds us, and when we participate in the food web, we feed the web of life.

Ecological science and ecological medicine recognise that the health of the gut microbiome is the basis of our health; most chronic diseases have their roots in the destruction of the gut microbiome. Healthy, biodiverse food grows a healthy gut and healthy food grows in healthy soils. Healthy soils are rich in organic matter and in soil biodiversity. Biodiversity intensive, photosynthesis intensive, regenerative organic farming draws out more carbon dioxide from the atmosphere, thus following nature's path of cooling the planet. One-third of the carbon fixed by the plant is returned to the soil as a soil exudate. Organic soils, rich in soil organism biodiversity, also add to the nutritional density and diversity of food, contributing to our health and nutrition.

Mycorrhizal fungi can hold 30 percent of emissions. Earthworms are important drivers of global food production, contributing to approximately 6.5 percent of grain yield. They contribute to soil health and resilience to climate change, as well. Soils with earthworms drain four to ten times faster than those without, and their water-holding capacity is higher by 20 percent. Earthworm castings, which can amount to four to thirty-six tons per acre, contain five times more nitrogen, seven times more phosphorous, three times more exchangeable magnesium, eleven times more potash, and one-and-a-half times more calcium. Their work on the soil promotes the microbial activity essential to living soil.

Building the infrastructure for oil has been the human preoccupation for at least one century. Cocreating the infrastructure of life with the earth and her beings must now become our commitment for the next century. In many cultures, traditional scientific knowledge recognised the links between ecology, agriculture, food and health, something that mechanistic science totally set aside.

We need to reconnect earth justice to human rights, to recognise the pain of the earth as connected to the pain of people. It is time to connect the climate crisis and the biodiversity crisis to the industrial food system. It is time to see that the same fossil fuel-based, chemical intensive, resource intensive, ultra-processed food systems that cause metabolic disorders for humans are leading to the metabolic disorder of the earth, whose symptom is climate change. At the root of the polycrisis is a mechanical, militaristic mind, a monoculture of the mind, which reduces the biodiverse, self-organised, living earth to raw material for the money machine. It is time to recognise the difference between the fake science and false solutions of the 1% and the deep ecological sciences of living systems, and real ecological solutions to the real, interconnected crises we face.

A paradigm shift requires walking a path beyond climate colonialism and climate change denial. It means walking the path of regenerating the earth as members of the earth family, interconnected and entangled in a thriving, living web of life. It means seeking climate justice and food freedom in our everyday lives, everywhere—reclaiming our food, reclaiming the earth, reclaiming our lives, our freedoms and our futures.

2 | Two Paradigms
Mechanophilia vs. Nature's Technology

The mechanistic paradigm is based on two assumptions: *terra nullius*, or dead earth; two, that life is made up of separable, immutable parts and can be understood through a fragmented and reductionist approach to each part, without recognising their mutual relationship and interaction. Each part can also be manipulated mechanically, without an adverse effect on the organism or the system, and without ecological consequence. Such a paradigm is based on separation, division, fragmentation, atomisation and the absolutism of fixed, unchanging properties.

The scientist and scholar Predrag B. Slijepčević repeatedly reminds us in *Biocivilisations: A New Look at the Science of Life* that 'life is neither mechanism nor thing. Instead, life is permanent change', that 'no organism is a machine'. He has called the mechanical philosophy 'mechanophilia, a love for the machine'.

While the mechanistic view is based on mastery and conquest over nature, the quantum paradigm, the ecological paradigm and the traditional knowledge systems of Indigenous peoples and others share an underlying understanding of an interconnected universe. Non-separability, inter-being, interconnectedness through diversity are the reality of our being and of nature.

The ecological paradigm which guides my scientific work is informed by the ancient science of agroecology, people's knowledge, as well as by my formal training in the non-mechanistic paradigm of quantum theory. Quantum theory shows us a world beyond the mechanistic assumptions of a natural world consisting of dead objects, immutable unchanging particles separated from each other with fully determined

Mechanophilia vs. Nature's Technology

and determinable coordinates such as mass, position, velocity and action, limited by contact and application of a force. The quantum world is not constituted of fixed particles, but of potential. A quantum can be a wave or a particle. It is indeterminate, and therefore, uncertain. It is non-separable, non-local; consequently, action at a distance becomes possible, and contrary to the mechanistic ideal of nature-human separation, the observer 'creates' the observed. An interactive, interrelated world of participation becomes possible.

A mechanistic philosophy evolved hand in hand with empire-building. Itwas central to the project of establishing the 'empire of man over inferior creatures of God', as Robert Boyle, widely considered one of the founders of modern chemistry, argued, and over 'lesser creatures', including women, all non-western cultures and all non-human beings.

What is imposed as 'science', moulded by the mechanical mind, was based on the subjugation of nature, women, diverse cultures and ordinary human beings. The American philosopher Sandra Harding has called it a 'Western, bourgeois, masculine' project, and according to physicist Evelyn Fox Keller,

> Science has been produced by a particular subset of the human race—that is, almost entirely by white, middle-class men. . . . For the founding fathers of modern science, the reliance on the language of gender was explicit: They sought a philosophy that deserved to be called 'masculine', that could be distinguished from its ineffective predecessors by its 'virile' powers, its capacity to bind nature to man's service and make her his slave.[1]

---

[1] Evelyn Fox Keller, *Reflections on Gender and Science* (New Haven: Yale University Press, 1985).

The mechanical mind is a construct of capitalist patriarchy and an instrument of the colonising empire. It is an efficient tool for exploitation and extraction, for manipulation and control, but is clumsy and ignorant for maintaining, rejuvenating, nourishing and growing life.

This epistemic violence is now being combined with the violence of corporate interests to viciously attack all scientific traditions, including those that have evolved from within western science and have, through autopoietic epistemic evolution, transcended the mechanistic worldview.

Diversity, balance, resilience and symbiosis are missing in the mechanistic framework. So, the most important aspects of living systems and living processes, and their interconnectedness, find no place in the mechanistic science of industrial agriculture and industrial medicine. This pathology of epistemology, as stated by the anthropologist Gregory Bateson, eroded the ontology of a living, intelligent earth and living, autopoietic organisms.

Living systems maintain balance and harmony and have the intelligence to return to a stable state of balance after a disturbance that creates imbalance. A system is autopoietic when its function is geared towards self-renewal; an autopoietic system refers to itself. Living systems are organised and regulated from within—internally, not externally.

In contrast, an allopoietic system, such as a machine, is assembled from outside, with external inputs. A machine refers to a function given from the outside, such as the production of a specific output. Life is self-organised, unlike a machine which is assembled externally.

The capacity to self-organise is the distinctive feature of living systems. They are autonomous. This does not mean that they are isolated and non-interactive; rather, such organisms

Mechanophilia vs. Nature's Technology

organise and regulate themselves by exchanging matter, energy and information with their surroundings. They interact with their environment but retain their autonomy, distinctiveness and uniqueness. The environment merely triggers the structural changes; it does not specify or direct them. The living system specifies its own structural changes and those patterns in the environment that will trigger them. A self-organising system knows what it needs to import and export in order to maintain itself.

Living systems are complex; the complexity in their structure allows for the emergence of new properties. One of the distinguishing properties of living systems is their ability to undergo continual structural change while preserving their form and pattern of organisation.

Living systems are also diverse; their diversity and unique attributes are maintained through spontaneous self-organisation. The components of a living system are continually renewed and recycled via structural interaction with the environment, yet the system maintains its organisation, as well as its distinctive form. Self-healing and repair are the other characteristics of living systems that derive from complexity and self-organisation.

Self-organising systems can heal themselves and adapt to changing environmental conditions; mechanical systems do not heal or adapt—they break down. When an organism or a system is mechanically manipulated to improve a one-dimensional function, either the organism's immunity decreases or becomes vulnerable to disease, or it becomes dominant in an ecosystem.

Self-regulating systems have the metabolic potential to return to a homeostatic state of balance. This is why living systems—the earth as a living being and the human body—can be healed. This requires negative feedback loops, which send feedback to the system to neutralise the harm and restore balance, that retain the conditions of living systems within the boundaries that support life. Destroying the negative feedback loop leads to tipping points and collapse. A negative feedback loop is a system-generated message that prevents it from continuing on a destructive path.

Thus, there are two paradigms for thinking of ourselves in the world and of our relationship with the earth. We can either think of ourselves as being separate from nature, or as being a part of it; and we can either think of nature as alive, self-organised, self-regulated, or as dead matter, raw material for industrial production. These two paradigms shape the two systems of agriculture—the ecological and the industrial—and they also shape the discourse around climate, food and health.

The first paradigm is ecological, based on interrelationships within nature and within our bodies, and between nature and humans. This ecological paradigm is shared by the ancient science of agroecology as well as by newer sciences like quantum theory and new findings in biological sciences about living systems—soil, plants, food, nutrition and health. This paradigm recognises intelligence in all life at all levels, from microbes and cells to our bodies and the planet earth. It sees ecological degradation and disease as impairment in the capacity for self-organisation and self-regulation, for healing and renewal. In the ecological paradigm, agriculture, food production and health are internal input systems with the capacity and potential to produce what they need to heal and repair. The earth is a living system. Food is a living system. Our bodies are living systems.

Interconnectedness and non-separability, self-organisation and self-regulation, potential and process, and complex systems

Mechanophilia vs. Nature's Technology

causality are the basis of transformation and change in living systems. The ecological paradigm is based on potential and process, not on immutable, unchanging entities.

The second paradigm is mechanistic and reductionist, based on seeing nature and our bodies as constituted of separate and disconnected parts and dead or inert matter. Nature, food, even our bodies are viewed as 'machines', managed externally with external inputs, external control and external regulation. Biodiversity and living beings are seen as 'objects' to be controlled and manipulated for extraction. Blindness to interconnectedness and symbiotic relationships leads to being blind to living processes and to 'action at a distance'.

In the reductionist paradigm, transformation and change require external force, and causality is Newtonian—linear, mechanical and reducible to mass and force. Thus, in the mechanistic paradigm, industrial agriculture is conceived as an external input system, based on buying expensive patented seeds and toxic agrichemicals; health is viewed as an external input system based on the purchase of expensive patented pharmaceuticals, additives and 'fortification'; and food is seen as 'mass', which can be manufactured, manipulated, substituted and engineered.

Goethe emphasises, 'Life as a whole expresses itself as a force that is not to be contained within any one part. . . . The things we call the parts in every living being are so inseparable from the whole that they may be understood only in and with the whole.'

In the first phase of the industrialisation of our food system, the complex soil web was substituted by nitrogen fertilisers made from fossil fuels. The slogan coined was 'bread from air'. Synthetic fertilisers destroyed the living soil, exhausted and polluted the waters, emitted a GHG three times more damaging to climate systems than carbon dioxide, replaced biodiversity with monocultures and destroyed the nutrition in food. The industrialising of agriculture and its output reduced the system to measuring the yields of monocultured commodities, discounting the quality of food, biodiverse outputs, the cost of external inputs and the ecological health of the system.

Mechanistic nutrition reduces food to its nutrient and biochemical constituent parts, which Gyorgy Scrinis, Associate Professor of Food Politics and Policy at the University of Melbourne, and Marion Nestle, molecular biologist and nutritionist, have referred to as 'nutritionism', a mere listing of constituents, ignoring the quality, the source and the process through which food is produced. This is nutritional reductionism. Scrinis says,

> I refer to this nutritionally reductive approach to food as the ideology or paradigm of nutritionism. This focus on nutrients has come to dominate, to undermine, and to replace other ways of engaging with food and of contextualising the relationship between food and the body.[2]

In the second phase of industrialisation, seeds were genetically engineered in order to patent them. While claiming patents, genetically modified organisms (GMOs) were defined as 'novel', as not having existed before; and while trying to escape responsibility for biosafety, they were unscientifically defined as 'substantially equivalent' to non-GMO seeds and food, leading to 'don't look, don't see, don't find the impacts and declare safe'.

---

[2] Gyorgy Scrinis, *Nutritionism: The Science and Politics of Dietary Advice* (New York: Columbia University Press, 2013).

Mechanophilia vs. Nature's Technology

In the current phase, fake food is being promoted as equivalent to real food, as a false solution to the climate crisis. When food is seen through the lens of nutritional reductionism or genetic reductionism, causation is artificially reduced to one cause and one effect, with both cause and effect being decontextualised.

In living systems, causality is systems causality, process causality and contextual causality. Properties and behaviours are potentials, and their expression depends on the context, on the relationship, on living processes, on complexity. As the scientist and author Giulia Enders writes in *Gut*,

> The important thing is not to reduce the human body to a two-dimensional, cause-and-effect machine. The brain, the rest of the body, bacteria and the elements in our food all interact with each other in four dimensions. Striving to understand all these axes is surely the best way to improve our knowledge.[3]

Linear causality, on the one hand, disconnects climate, agriculture, food and health; on the other hand, it allows claims to be made linking specific tools to complex, multicausal phenomena. It shifts our gaze from complex systems to tools and technology without assessing how the tools impact the system. In the Green Revolution narrative, Norman Borlaug's 'miracle' dwarf wheat varieties, bred for chemicals, were supposed to have increased food production in India. But, as Navdanya's subsequent studies have shown, food is more than wheat and rice.[4] We need vegetables, pulses, oilseeds and millets

---

[3] Giulia Enders, *Gut: The Inside Story of Our Body's Most Underrated Organ* (London: Scribe Publications, 2015), 178.

[4] Vandana Shiva, *Agroecology and Regenerative Agriculture: Sustainable Solutions for Hunger, Poverty, and Climate Change* (Santa Fe: Synergetic Press,

for a balanced diet. An increase in rice and wheat monocultures displaced India's rich biodiversity and led to a decline in the production of diverse foods. Acreage given over to rice and wheat monocultures made for greatly increased irrigation. Land and water contributed to a far higher production of rice and wheat, *falsely associated with new seeds and chemicals*, and then falsely extrapolated to increased food, when, in fact, there was a marked decrease in the diversity of foods produced and consumed. Navdanya's conservation of biodiversity and research on nutrition[5] also shows that calling industrial varieties 'high yielding' is inaccurate, because in terms of nutrition, they are low yielding compared to indigenous varieties. Linear causality applied to complex systems allows corporations producing harmful chemicals and GMOs to falsely claim increased yields when there is actually a 'failure to yield', and to simultaneously deny the harmful impact of their products.

The impact of bad agricultural practices on destabilising the climate and creating disease, and the impact of biodiverse agriculture on contributing to regenerating the earth and our health, cannot be gauged from a mechanistic, reductionist perspective based on only *one* element—carbon—linked to destabilising earth systems and climate systems, and on *one* part of our diet, linked to *one* disease. Complex system interactions call for systems causality and contextual causality, not mechanical causality. In mechanical causality, one element, one particle, one gene, one molecule, one ingredient can cause one specific

---

2022); Vandana Shiva with Vaibhav Singh, *Health per Acre: Organic Solutions to Hunger and Malnutrition* (New Delhi: Navdanya/Research Foundation for Science, Technology and Ecology, 2011).

[5] Vandana Shiva with Vaibhav Singh, *Health per Acre*.

impact. In systems causality and contextual causality, a system *as a whole* leads to potentials and tendencies for multiple changes in another complex system.

When we grow food in accordance with ecological laws, we regenerate the earth, her soil and biodiversity, her climate system. When we produce and distribute food on the basis of fossil fuels and fossil chemicals, we destroy the complex, interconnected systems of soil, water, biodiversity and climate. When we eat food grown with care, in alignment with nature's laws, free of chemicals and ultra-processing, that food, a complex system, interacts with six trillion cells in our body, another complex system.

The climate crisis is a result of reducing the planet to a machine run by fossil fuels, and our lives and food systems run by fossil chemicals, petrochemicals and fossil fertilisers. It is a result of blindness to the self-organised nature of living systems, including the earth as a living organism. The mechanistic industrial paradigm does not have the epistemic or intellectual and scientific potential to understand the roots of the climate havoc or the disease epidemic it has created, nor does it offer lasting solutions to either.

A misdiagnosis

A mechanical perspective reduces the earth to a single substance—carbon. Climate systems, which are complex, are also being reduced to one element—dead carbon. Living carbon, which is the basis of life and is ecologically and ontologically very different from fossil carbon, is being falsely equated with, and reduced to, dead carbon. Nitrous oxide from synthetic fertilisers, three hundred times more climate-damaging than CO_2, does not enter the discussion. Water is left out, even though most climate disasters and deaths are related to extreme water events—floods and drought. The ontology of the earth is reduced to carbon; dead fossil carbon, which brings death and destruction, is falsely equated with living carbon, which is life. Similarly, the complexity and multi-dimensionality of climate systems and processes are being reduced to only one parameter: rising temperatures and the warming of the earth—global warming.

The two basic cycles of life are the nutrient cycle and the water cycle, and both are central to the food system and to the earth's metabolism. The nutrient cycle, in turn, is the food cycle; food and climate are deeply connected through plants and photosynthesis. The ecological crisis is a consequence of the disruption of these cycles.

Charles Eisenstein, who writes on environmentalism, economics and philosophy, says,

> Earth is best understood as a living being with a complex physiology, whose health depends on the health of her constituent organs. Her organs are the forests, the wetlands, the grasslands, the estuaries, the reefs, the apex predators, the keystone species, the soil, the insects, and indeed every intact ecosystem and every species on earth. If we continue to degrade them, drain them, cut them, poison them, pave them, and kill them, earth will die a death of a million cuts. She will die of organ failure—regardless of the levels of greenhouse gases. . . . The core of the crisis is not warming, it is ecocide— the killing of ecosystems, the killing of life.[6]

It is when we see the earth as a living system that we begin to see climate change as a metabolic disorder of the earth, as

---

[6] Charles Eisenstein, 'How the Environmental Movement Can Find Its Way Again', March 26, 2023, https://charleseisenstein.substack.com/p/how-the-environmental-movement-can.

a symptom of our dysfunctional relationship with her. This is what Amitav Ghosh has called 'The Great Derangement': 'Who can forget those moments when something that seems inanimate turns out to be vitally, even dangerously alive?'[7]

Nature's technologies

R. Buckminster Fuller, visionary architect, has said, 'Universe is nothing but incredible technology.' The 'intellectual integrity of an eternally regenerative universe' is how he has described its complexity. He insisted that the key principle was to *recognise* nature as technology:

> In its complexities of design integrity, the Universe is technology. The technology evolved by man is thus far amateurish compared to the elegance of nonhumanly contrived regeneration. Man does not spontaneously recognize technology other than his own, so he speaks of the rest as something he ignorantly calls Nature. The Natural is the real Technological.[8]

Photons absorbed from the sun's energy by green plants split water molecules and reduce carbon dioxide, resulting in the formation of carbohydrates and oxygen. This is the foundational ecological cycle that sustains life on earth. It supports many secondary and tertiary cycles and epicycles. This complex coupling constitutes the *metabolism* of living systems. The geneticist Mae-Wan Ho elaborates,

---

[7] Amitav Ghosh, *The Great Derangement: Climate Change and the Unthinkable* (Chicago: The University of Chicago Press, 2016).

[8] 'R. Buckminster Fuller: The Science of Design', Weird Tales Designs, June 23, 2023, https://www.weirdtalesdesigns.com/r-buckminster-fuller-the-science-of-design.

Metabolism refers to the totality of chemical reactions that make and break molecules, whereby the manifold energy transformations of living systems are accomplished. The secret of living metabolism—which has as yet no equal in the best physicochemical systems that scientists can now design—is that the energy-yielding reactions are always coupled to energy requiring reactions. The coupling can be so perfect that the efficiency of energy transfer is close to 100%. . . .

Coupled cycles are the ultimate wisdom of nature. They go on at all levels, from the ecological down to the molecular through a wide range of characteristic timescales from millennia to split seconds. Thus, the transformation of light energy into chemical energy by green plants yields food for other organisms whose growth and subsequent decay provide nutrients in the soil on which green plants depend. The energy in foodstuffs is transformed into the mechanical, osmotic, electrical and biosynthetic work both within plants themselves and in other organisms in all the trophic levels dependent on green plants. Each kind of energy transformation in organisms is carried out by its own particular troupe of busy molecular machines working in ceaseless cycles. And upon all of these turn the innumerable life cycles of multitudinous species that make up the geological cycles of the earth.[9]

The same processes that cooled the planet and allowed life to emerge in its diversity are the processes that sustain, maintain and regenerate life. Conserving and regenerating the earth's biodiversity is the most significant climate action and health action that we can undertake.

---

[9] Mae-Wan Ho, *The Rainbow and the Worm: The Physics of Organisms*, 3rd ed. (Singapore: World Scientific Publishing Co., 2008), 54–55, https://www.worldscientific.com/doi/suppl/10.1142/6928/suppl_file/6928_chap04.pdf.

Mechanophilia vs. Nature's Technology

Over four billion years, the earth evolved microbes and plants which, through the process of photosynthesis, cooled the earth by capturing the carbon dioxide in the atmosphere with the help of sunlight and energy from the sun. This is nature's sophisticated 'carbon capture' technology which allowed carbon recycling, transforming CO_2 to O_2. Oxygen accumulated in the atmosphere and the earth was transformed from the original heat-trapping, CO_2-rich atmosphere to the reduced CO_2 atmosphere through the oxidising process of plants and living organisms. This allowed temperatures to be regulated at levels that support human and other biological life on earth.

Four billion years ago, the earth was a hot, lifeless planet. Through evolution, the earth and her biodiversity reduced the carbon rich atmosphere of the planet from 4,000 ppm to 250 ppm; and her temperature from 290°C, without life, to 13°C, with biodiversity. *Terra madre* created the conditions for the evolution of life's diversity; 200,000 years ago, she created the conditions for our species to evolve. We are among the youngest siblings in the earth family. In an age of collapse, climate catastrophes and extinction, we need to turn to the earth and to our evolutionary elders—plants—to learn, once again, how to live sustainably on earth and sow the seeds of hope, the seeds of the future.

Slijepčević reminds us that we have only been around for 0.01 percent of the time. For 99.9 percent of the time, the planet has evolved without us. To learn how to deal with climate change, we need to attend to the earth and her biodiversity, which have created the conditions for our species to evolve, sustain ourselves and provide for the basic needs of food, clothing and shelter in partnership with the biosphere and other species. We are interbeings. We are made of microbes. We are made of the plants that give us food.

Food and energy are the currencies of life, and all three begin with the green leaf of the plant. Sir Albert Howard writes in *An Agricultural Testament*,

> The energy for the machinery of growth is derived from the sun; the chlorophyll in the green leaf is the mechanism by which this energy is intercepted; the plant is thereby enabled to manufacture food—to synthesize carbohydrates and proteins from the water and other substances taken up by the roots and the carbon dioxide of the atmosphere. The efficiency of the green leaf is therefore of supreme importance; on it depends the food supply of this planet, our well-being, and our activities. There is no alternative source of nutriment. Without sunlight and the green leaf our industries, our trade, and our possessions would soon be useless.[10]

The seed, with the blessing of the sun, grows into plants that become the green mantle of the earth, returning part of the plants to the soil as organic matter to create living soil. The very processes that regulate the climate also provide us with food. Climate and food are interconnected, because ecologically, all life is plant-based. Yet, we have been blind to the creative and regulatory processes in plants.

Ecology, energy, entropy

Regeneration begins with the consciousness that we are part of one interconnected entity, the earth, rich in biodiversity. Sickness

---

[10] Sir Albert Howard, *An Agricultural Testament* (Oxford University Press, 1921), 23.

Mechanophilia vs. Nature's Technology

lies in broken processes and relationships, while the potential for health lies in diverse relationships which create, maintain and regenerate health. Mae-Wan Ho reminds us that,

> Life uses the highest grade of energy and organises it. It does not 'heat up' the body of the organism. Living systems can do this by their meticulous space-time organisation, in which energy is stored in a range of time-scales and spatial extents.

Unlike mechanical energy that runs mechanical systems, the living energy of living systems is based on negative entropy. In *What is Life?*, Schrödinger observes,

> It is by avoiding the rapid decay into the inert state of 'equilibrium' that an organism appears so enigmatic. What an organism feeds on is negative entropy.... Or, to put it less paradoxically, the essential thing in metabolism is that the organism succeeds in freeing itself from all the entropy it cannot help producing while alive.[11]

Living systems increase the organisation of life through negative entropy. Mechanical, industrial systems increase entropy, or disorder. Entropy indicates the waste and pollution created by mechanical systems. Construed in terms of order, the entropy principle tells us that the natural development of any total system is towards states of greater disorder.

According to the Hungarian biochemist and Nobel laureate Albert Szent-Györgyi de Nagyrápolt,

> It is common knowledge that the ultimate source of all our energy and negative entropy is the radiation of the sun. When

---

[11] Erwin Schrödinger, *What is Life? With "Mind and Matter" and "Autobiographical Sketches"* (Cambridge University Press, 1992).

a photon interacts with a material particle on our globe it lifts one electron from an electron pair to a higher level. This excited state as a rule has but a short lifetime and the electron drops back within 10^{-7} to 10^{-8} seconds [less than 100 millionth of a second] to the ground state, giving off its excess energy in one way or another. Life has learned to catch the electron in the excited state, uncouple it from its partner and let it drop back to the ground state through its biological machinery utilising its excess energy for life processes.[12]

Mae-Wan Ho clarifies,

> The biosphere . . . does not make its living by absorbing heat from the environment. No organism can live like a heat engine. . . . Instead, life depends on catching an excited electron quite precisely—by means of specific light absorbing pigments—and then tapping off its energy as it falls back towards the ground state. Life uses the highest grade of energy, the packet or quantum size of which is sufficient to cause specific motion of electrons in the outer orbitals of molecules. It is on account of this that living systems can populate their high energy levels without heating up the body excessively, and hence contribute to what Schrödinger intuitively identifies as 'negative entropy'. But what enables living systems to do so? It is none other than their meticulous space-time organisation in which energy is *stored*. . . . Energy flow organises the system, which in turn organises the energy flow.[13]

Ecological systems of food production and consumption are autopoietic, based on endosomatic energy or energy generated

---

[12] Albert Szent-Györgyi, in W.D. McElroy and B. Glass, eds., *Symposium on Light and Life* (Baltimore: Johns Hopkins Press, 1961), quoted in Mae-Wan Ho, op. cit., 57.

[13] Mae-Wan Ho, op. cit., 70.

by the living system. Economic systems based on ecological production and on basic needs aim at low entropy. Economic systems based on industrial production and on maximising profits are high entropy systems. The irreversible flow of energy and matter, which is implied by an entropic transformation, led the Romanian American mathematician and economist Nicholas Georgescu-Roegen to assert that 'entropy is the sole temporal law in physics'.[14] In his book *The Entropy Law and the Economic Process*, he writes,

> It is natural that the appearance of pollution should have taken by surprise an economic science which has delighted in playing around with all kinds of mechanistic models. Curiously, even after the event economics gives no signs of acknowledging the role of natural resources in the economic process. Economists still do not seem to realize that, since the product of the economic process is waste, waste is an inevitable result of that process and *ceteris paribus* increases in greater proportion than the intensity of economic activity.[15]

Mae-Wan Ho points out that

> A living system differs from a conventional thermodynamic machine. A living system is a much more complex and dynamic organisation that somehow empowers it to metabolise, grow, differentiate, and maintain its individuality and vibrant wholeness, something no physical system can yet do.[16]

Thermodynamics has its origin in describing the transformation of heat energy into mechanical work. Concepts of

---

[14] Nicholas Georgescu-Roegen, *The Entropy Law and the Economic Process*, 1st ed. (Cambridge, MA: Harvard University Press, 1971, 141.

[15] Georgescu-Roegen, *The Entropy Law*, 19.

[16] Mae-Wan Ho, op. cit.

thermodynamics are a result of the fossil fuel industrial age, when attempts were being made to determine how much mechanical work can be made available from steam engines fuelled by coal. Fossil fuel addiction has created a way of thinking that I call 'fossilised'. It is a monoculture of the mind, blind to biodiversity and its regenerative potential. It has created a fossil fuel-dependent industrial mode of production for meeting our daily needs. We eat oil. We drink oil. We breathe oil. The fossil age has created government policies and economic policies which privilege oil and oil-based systems and punish soil and soil-based local economies. The last two centuries of dependence on fossil fuels have created multiple distortions in our view of our production and consumption systems, our ideas of efficiency and productivity, our bias towards technological progress and of the way we produce and distribute our food. We use more resources to produce the goods we consume, and we call this being more 'productive'. We create more waste and more externalities that the earth and others have to bear, and we call this being more 'efficient'. We degrade the planet, push species to extinction, make the planet unliveable because of climate chaos, and call this 'progress'.

3 | Industrial Farming and the Illusion of Food Security

Nature created coal and oil by fossilising the living carbon of plants and other organisms over six hundred million years. In a short two hundred years, we learned to mine and burn the biological matter that the earth had fossilised. She left it underground so that we could leave it underground. Instead, we mined and mined to extract coal and oil, destroy the earth, her forests and farms, her soil and water, her climate systems and biodiversity, chasing the illusion that creating the heavy polluting infrastructure of oil was an indicator of progress.

The fossilised paradigm has created four illusions. The first is the illusion of separation and mastery—that we are separate from the earth, are her owners, masters and manipulators. I refer to this illusion as *ecological apartheid*. The German scientist Justus von Liebig referred to it as the *metabolic rift*. The second illusion is that the earth represents raw material, which derives value only through extraction and exploitation. The extraction of fossil fuel created the entire edifice of extractive technologies and extractive economies. The third illusion is anthropocentric: that the earth's resources are intended only for humans and that humans are superior to other species which are to be owned, manipulated, exploited, engineered and made expendable. The fourth illusion is that indigenous ecological cultures and civilisations, which are non-industrial and fossil fuel-free, are backward and primitive because they create and sustain the infrastructure of life based on co-creation and co-production. Indigenous cultures are sophisticated in terms of their ability to live in harmony with biodiversity—80 percent biodiversity is on the 20 percent land where Indigenous peoples live.[1]

Development was defined in the Age of Oil as the use of fossil fuels and petrochemicals, of plastic and pesticides derived from oil. The pollution from the Age of Oil is everywhere for us to see. The disruption of the climate system is not as visible a form of pollution as plastic waste, nor is the pollution of our food with pesticides and toxic agrichemicals.

Non-violent, biodiverse, living carbon economies and the technologies of biodiversity in indigenous cultures have been violently displaced by dead carbon economies and technologies of fossil fuels and their toxic derivatives. Indigenous peoples knew of fossil deposits before the Age of Oil and used them in tiny quantities. Across the world, local communities are now pitted against the destructive impact of the extractive industry of coal and oil, as well as the toxics and poisons that are derived from petrochemicals. The practices and actions that drive climate change have led to biodiversity loss and species extinction; they have also created the hunger, malnutrition and health emergencies.

Industrial agriculture and globalised food systems

The earth has planetary boundaries; its most seriously ruptured boundaries are those of biodiversity/genetic diversity and

---

[1] Anna Fleck, 'Indigenous Communities Protect 80% of All Biodiversity', Statista, July 19, 2022, https://www.statista.com/chart/27805/indigenous-communities-protect-biodiversity/; see also Convention on Biological Diversity, 'Indigenous Communities Protect 80% of All Biodiversity', July 20, 2022, https://www.cbd.int/kb/record/newsHeadlines/135368.

nitrogen. The erosion of genetic diversity and the transgression of the nitrogen boundary have already crossed catastrophic levels. Industrial chemical agriculture is based on external inputs of nitrogen, phosphorous and potassium, and on industrial monocultures of globally traded commodities.

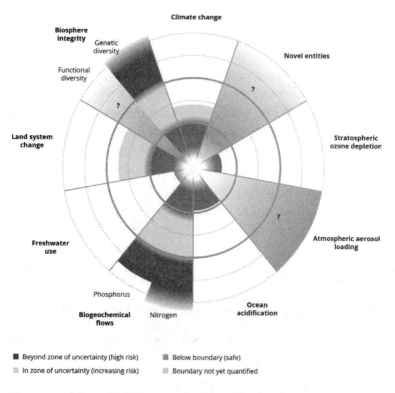

Estimates of how the different control variables for planetary boundaries have changed from 1950 to the present.

(Source: W. Steffen et al., 'Planetary Boundaries: Guiding Human Development on a Changing Planet', Science 347, no. 6223 (January 2015), cited in Shiva, et al., Seeds of Hope, Seeds of Resilience [New Delhi: Navdanya/RFSTE, 2017], 5, https://navdanyainternational .org/publications/seeds-of-hope-seeds-of- resilience.)

Industrial Farming and the Illusion of Food Security

Industrial monocultures are an important driver of destruction and erosion of biodiversity, both in forests and on farms. The Amazon and Indonesian rainforests are being destroyed due to the monocultures of Roundup Ready soyabean and palm oil.

- Deforestation by agribusiness contributes to 20 percent of GHGs.
- Land use change and deforestation, including GMO soya, are reducing the Amazon rainforest by 15–20 percent.
- Processing, transport, packaging and retail of industrial food is responsible for 15–20 percent of GHGs.
- Waste emits 2–4 percent of GHGs.

From the way food is grown today to the way it is processed and then distributed, every step has a bearing on climate change. The pie chart below shows the extent of the impact made by each stage of industrial food production.

Chemical intensive agriculture contributes between 11 to 15 percent of GHG emissions, of which nitrous oxide, released by

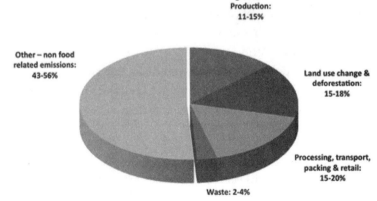

Food and climate change

(*Source*: GRAIN)

synthetic fertilisers, is the main culprit. Artificial fertilisers are made by burning fossil fuels at high temperatures to fix atmospheric nitrogen through what is called the Haber–Bosch process.

As the second segment of the pie chart shows, land use change, especially deforestation, is another factor that greatly impacts climate change, contributing between 15 to 18 percent of GHG emissions. Biodiverse and sustainable farms have switched to chemical monoculture cultivation, and forests are being cut down to cultivate palm and soyabean or to become grazing grounds for cattle to provide meat for the fast-food industry.

The industrial processing and ultra-processing of food not only robs it of vital nutrients and its subtle thermal qualities but also majorly aggravates the climate situation as it spews carbon dioxide into the atmosphere through the use of fossil fuel as a source of energy. Further, when food has travelled thousands of miles, by road, air or rail, it contributes neither to your health nor to the planet's. Since supermarkets and mega retail outlets require long shelving time, all products are heavily packaged, which again aggravates the situation.

Food waste, caused by long-distance transportation and globalised food chain production, privilege uniformity; 50 percent of the food in North America is wasted in its journey from the farm to the table. America throws away nearly half its food.[2]

Fossil food systems are toxic systems, based on technologies of extermination and death. Synthetic fertilisers kill soil organisms, pesticides and insecticides kill insects, herbicides kill

---

[2] Jonathan Bloom, *American Wasteland: How America Throws Away Nearly Half of Its Food (and What We Can Do About It)* (Boston: Da Capo Press, 2010).

[3] Katy Daigle and Julia Janicki, 'Extinction Crisis Puts 1 Million Species on the Brink', Reuters, December 23, 2022, https://www.reuters.com/lifestyle/science/extinction-crisis-puts-1-million-species-brink-2022-12-23.

Industrial Farming and the Illusion of Food Security

plants. One million species are threatened with extinction; 200 disappear every day.³

Rachel Carson, author of the prophetic *Silent Spring*, claimed that at the heart of our motivation to introduce poisons into the environment lies a deeply-held and outdated philosophy: 'The "control of nature" is a phrase conceived in arrogance, born of the Neanderthal age of biology and philosophy when it was supposed that nature exists for the convenience of man.'⁴

According to the Inter-Governmental Panel on Biodiversity and Ecosystem Services (IPBES),

> Rapid expansion and unsustainable management of croplands and grazing lands is the most extensive global direct driver of land degradation, causing significant loss of biodiversity and ecosystem services—food security, water purification, the provision of energy and other contributions of nature essential to people. This has reached 'critical' levels in many parts of the world.

Professor Robert Scholes from South Africa, co-chair of this assessment, with Dr Luca Montanarella of Italy, warned, 'With negative impacts on the well-being of at least 3.2 billion people, the degradation of the Earth's land surface through human activities is pushing the planet towards a sixth mass species extinction...'⁵

According to the International Union for Conservation of Nature (IUCN), over-exploitation and agriculture are the 'big

---

⁴ Brian Payton, 'Rachel Carson (1907–1964)', Nasa Earth Observatory, November 13, 2002, https://earthobservatory.nasa.gov/features/Carson/Carson2.php.

⁵ 'Worsening Worldwide Land Degradation Now "Critical", Undermining Well-Being of 3.2 Billion People', IPBES, March 26, 2018, https://www.ipbes.net/news/media-release-worsening-worldwide-land-degradation-now-'critical'-undermining- well-being-32.

killers' with the greatest current impact on biodiversity.[6] A 2017 German study has shown that 75 percent of its insects have disappeared.[7] Another study in 2018 from France has called the disappearance of birds in France a 'biodiversity oblivion'.[8] Yet another study on the effects of pesticides, published in 2023, found there are around 800 million fewer birds in Europe now than there were 40 years ago, a drop of 25 percent. In wild farmland birds, which rely on insects for food, the decline is 60 percent.[9]

BirdLife International's authoritative report *State of the World's Birds 2022* estimates that there are now nearly three billion fewer wild birds in Canada and the US than a few decades ago.[10]

---

[6] Monique Grooten and Rosamunde Almond, eds., *Living Planet Report—2018: Aiming Higher* (Switzerland: WWF, 2018), 28, https://www.worldwildlife.org/pages/living-planet-report-2018.

[7] 'Insects Decline Dramatically in German Nature Reserves: Study', Phys.org, October 18, 2017, https://phys.org/news/2017-10-three-quarters-total-insect-population-lost.html.

[8] 'France's Bird Population Collapses Due to Pesticides', Return to Now, March 25, 2018, https://returntonow.net/2018/03/25/frances-bird-population-collapses-due-to-pesticides.

[9] Polly Dennison, 'Farming Linked to Decline in Birds', LinkedIn News, May 17, 2023, https://www.linkedin.com/news/story/farming-linked-to-decline-in-birds-5648012; see also Perrine Mouterde and Stéphane Foucart, 'Pesticides and Fertilizers Are Driving the Decline of European Bird Populations', Le Monde, May 16, 2023, https://www.lemonde.fr/en/environment/article/2023/05/16/pesticides-and-fertilizers-are-driving-the-decline-of-european-bird-populations_6026804_114.html.

[10] *State of the World's Birds 2022*, BirdLife International, https://www.birdlife.org/papers-reports/state-of-the-worlds-birds-2022; see also Ian Angus, 'Industrial Farming Has Killed Billions of Birds', Climate & Capitalism, June 6, 2023, https://climateandcapitalism.com/2023/06/06/industrial-farming-kills-billions-of-birds.

Industrial Farming and the Illusion of Food Security

Farmland practices are also driving a bird population decline across Europe.[11] Researchers have found that increased farm sizes resulted in a 15 percent decline in bird diversity in the region.[12]

The interrelated aspects of the ecological crisis are creating new vulnerabilities for food and farming. According to WWF's 2018 *Living Planet Report*, since 1970, when industrial agriculture and chemicals began spreading, we have wiped out 60 percent of the animals on the planet, and freshwater species declined by 83 percent over the same period. Since 1960, the global ecological footprint has increased by more than 190 percent! Globally, the extent of wetlands was estimated to have declined by 87 percent since 1970.[13] 'We are sleepwalking towards the edge of a cliff,' warns WWF's Mike Barrett.[14]

The above evidence indicates how, in spite of its vital importance for human survival, biodiversity is being lost at an alarming rate, as hundreds of species disappear daily with the spread of chemical intensive, capital intensive agriculture. This poison-based, monoculture-based agriculture is wiping out the huge diversity of crops we once grew and ate; humans

---

[11] Stanislas Rigal et al., 'Farmland Practices Are Driving Bird Population Decline Across Europe', *PNAS* 120, no. 21 (May 2023), https://www.pnas.org/doi/10.1073/pnas.2216573120.

[12] 'New Study Shows the Toll Industrial Farming Takes on Bird Diversity', ScienceDaily, January 12, 2022, https://www.sciencedaily.com/releases/2022/01/220112154933.htm.

[13] Monique Grooten and Rosamunde Almond, eds., *Living Planet Report—2018: Aiming Higher* (Switzerland: WWF, 2018), 7, https://www.wwf.org.uk/updates/living-planet-report-2018.

[14] Julia Conley, 'Humanity "Sleepwalking Towards the Edge of a Cliff": 60% of Earth's Wildlife Wiped Out Since 1970', Common Dreams, October 30, 2018, https://www.commondreams.org/news/2018/10/30/humanity-sleepwalking-towards-edge-cliff-60-earths-wildlife-wiped-out-1970.

consumed more than 10,000 species of plants before globalised industrial agriculture became widespread.

The commodification of food has reduced the crops cultivated to a dozen globally traded commodities.[15] In India, we evolved 200,000 rice varieties, 1,500 mango and banana varieties and 4,500 varieties of brinjal. We bred our crops for diversity and nutrition, taste, quality and climate resilience. Today, we are growing a handful of chemically grown commodities which are nutritionally empty and laden with toxics; 93 percent of our crop diversity has been pushed to extinction through industrial agriculture. We are facing a severe crisis of malnutrition—India is 99th on the hunger index.

As the Leipzig Conference on Plant Genetic Resources acknowledged in 1996, 75 percent of the disappearance of agricultural biodiversity was due to the introduction of 'modern varieties'. It was also recognised that 'the imposition of WTO and the globalisation of industrial agriculture has led to further acceleration of the disappearance of diversity'.[16]

It was in Leipzig that the feminist scholar Maria Mies and I issued the Leipzig Appeal to say 'No to GMOs and No to Patents on Seed'.

The Leipzig Appeal, written almost three decades ago, is as relevant today as it was then. Though many initiatives have been taken, the world over, to reclaim Food Security, Food Safety and Food Sovereignty at individual and community levels, the industrial food system continues unabated to impose its dystopia.

---

[15] *The Law of the Seed*, Navdanya International, May 20, 2013, https://navdanyainternational.org/publications/the-law-of-the-seed.

[16] Food and Agriculture Organization, 'International Conference on Plant Genetic Resources to Meet at Leipzig from 17 to 23 June', United Nations, June 13, 1996, https://press.un.org/en/1996/19960613.fao3635.html.

> ### Leipzig Appeal for Women's Food Security[17]
>
> As a preparation to the World Food Summit, a FAO conference on Plant Genetic Resources took place in Leipzig, Germany, in June 1996. An independent NGO meeting preceded it, entitled 'In Safe Hands: Communities Safeguard Biodiversity and Food Security'. At this conference, several women from the South and North observed that the whole discussion on food security did not take into account the fact that it is women worldwide who provide most food, both as producers and as consumers, to their families and their communities.
>
> They decided to formulate a statement rejecting the trend to remove food security from the hands of communities and farmers, and to criticise the neoliberal policy of global food trade and the genetic manipulation of food for the sake of profit. They felt the need to present a critique from a women's perspective on the proposed policy of globalisation, liberalisation and industrialisation of food production, trade and consumption. Therefore, an appeal was formulated and signed by Prof Maria Mies and Dr Vandana Shiva called 'The Leipzig Appeal for Women's Food Security' which was first circulated at the Leipzig Conference and later on. . . around the world. It received hundreds of signatures in support of this position.

---

[17] Maria Mies and Vandana Shiva, 'Leipzig Appeal for Women's Food Security', June 20, 1996, https://www.iatp.org/sites/default/files/Leipzig_Appeal_for_Womens_Food_Security.htm.

Food Security in Women's Hands: Food Sovereignty for All, No to Novel Food and No to Patents on Life

For thousands of years, women have produced their own food and guaranteed food security for their children and communities. Even today, 80 percent of the work in local food production in Africa is done by women. In Asia, 50 to 60 percent and in Latin America, 30 to 40 percent. And everywhere in the world, women are responsible for food security at the household level. In patriarchal societies, however, this work has been devalued.

Historically, all societies have survived because they provide food security to their people. This policy, however, has been subverted by the globalisation, trade liberalisation, industrialisation and commercialisation of all agricultural products under the auspices of the World Trade Organization (WTO) and the World Bank/International Monetary Fund (IMF).

In November 1996, the UN's Food and Agriculture Organization will hold a World Food Summit in Rome. Its goal is to achieve 'universal food security' by the year 2020, eradicating hunger and malnutrition. However, the technical preparatory papers show that this objective is to be met through a continuation and extension of industrialisation and the worldwide trade in food. Food will be produced where labour is cheapest and environmental protections weakest. Poor communities will be forced to produce luxury products for export to rich countries and classes. These trends are already in effect, with devastating results: the large-scale disappearance of small farmers, the end of food self-sufficiency, reliance on monocultures, genetic manipulation of food, loss of biodiversity and ecological sustainability. The impoverished rural people who are displaced through this world agriculture policy end up as marginal members of society in overcrowded megacities without work, hope or food. Although it is known that this policy is the cause of poverty and malnutrition, it is still proposed as a remedy for these very ills. The most vulnerable groups affected by these policies are poor rural women and children.

This policy also threatens food safety in the North, where the family farm has been rapidly replaced by chemical intensive agribusiness. Consumers have become virtual hostages to a handful of transnational food processing and trading corporations. At the consumption end of the globalised food chain, women as housewives can no longer guarantee that they can give their families wholesome and healthy food.

In Peru, Chile, and other countries of the South, women are fighting against this monopolistic policy, building their own communal food and health systems. Women in indigenous societies fight against land alienation; women in export-oriented agriculture oppose hazardous chemicals. They are supported by women in the North who call for boycotts of these export products: flowers, vegetables, shrimps. Many groups in the North and South reject genetic manipulation of food. We are told that this biotechnology is necessary to feed a growing world population. However,

60 percent of cereals are fed to animals in industrial farming systems. More and more land in the South is not used for nourishing local people, but for the production of luxury items for export.

The commercial interests connected with this technology are particularly apparent in the promotion of patenting of life-forms—plants, animals and humans—under the protection of Trade Related Intellectual Property Rights (TRIPs). In the South, the patenting of life-forms is opposed because it is, in many cases, based on simple piracy: theft of indigenous biodiversity and local knowledge. In the North, many people oppose patents on life-forms for ethical reasons.

On the consumer side, a majority of Europeans oppose genetically manipulated foods. Yet the European Union promotes such 'novel food', even refusing to label it, thus denying consumers their human and civil right to determine what they eat. Consumption in this so-called 'free market' becomes a matter of coercion.

Worldwide, women are resisting the policies which destroy the basis of their livelihood and food sovereignty. They also create alternatives to guarantee food security for their communities based on different principles and methods than those governing the dominant, profit-oriented global economy. They are:
- localisation and regionalisation instead of globalisation
- non-violence instead of aggressive domination
- equity and reciprocity instead of competition
- respect for the integrity of nature and her species

- understanding humans as part of nature instead of as masters over nature
- protection of biodiversity in production and consumption

Food security for all is not possible within a global market system based on the dogma of free trade, permanent growth, comparative advantage, competition and profit maximisation.

On the other hand, food security can be achieved if people within their local and regional economies feel responsible, both as producers and as consumers, for the ecological conditions of food production, distribution and consumption, and for the preservation of cultural and biological diversity where self-sufficiency is the main economic goal.

Our food security is too vital an issue to be left in the hands of a few transnational corporations with their profit motives, or up to national governments that increasingly lose control over food security decisions or to a few—mostly male—national delegates at UN conferences who make decisions affecting all our lives.

Food security must remain in women's hands everywhere! And men must share the necessary work, be it paid or unpaid. We have a right to know what we eat! No to Novel Food and No to Patents on Life. We will resist those who force us to produce and consume in ways that destroy nature and ourselves!

Maria Mies and Vandana Shiva

Climate change, desertification and the water crisis

Synthetic nitrogen fertiliser is the basis of fossil fuel-based industrial agriculture. The Center for International Environmental Law (CIEL) has linked it to 'laundering fossil fuels in agrichemicals'. Ecological agriculture regenerates the living soil through the circular economy of what Sir Albert Howard referred to as the 'law of return', of giving back to the earth part of what the earth gives us. Liebig, the founder of modern organic chemistry, called it the 'law of recycling'.

Commercial interests transformed soil fertility into external inputs. Initially, trade in Guano[18] from Chile was promoted for inputs in nitrogen fertiliser; eleven million tonnes of Guano were used up in less than 20 years. That is when two scientists from IG Farben (the cartel that worked for Hitler), Fritz Haber and Carl Bosch, developed the Haber–Bosch process of heating fossil fuels at high temperatures of 400–500°C under 200–400 atmospheres of pressure. This allows the double bond of nitrogen in the air to split and join the hydrogen from the gas, creating ammonia, which becomes liquid when cooled. From liquid ammonia, ammonium nitrate is produced which is then used in bombs and explosives. The same process also produces nitrogen fertiliser.

Giant factories producing fertiliser based on the Haber–Bosch process 'drink rivers of water, inhale oceans of air' and burn 1 percent of the earth's energy. The manufacture of synthetic

---

[18] Guano is the accumulated excrement of seabirds or bats. It is a highly effective fertiliser due to its high amounts of nitrogen, phosphate and potassium, all key nutrients essential for plant growth. Guano was also sought for the production of gunpowder and other explosive materials.

Industrial Farming and the Illusion of Food Security

fertilisers was likened to 'pulling bread from air'.[19] The reality that was hidden by this description was the fact that synthetic fertilisers are fossil fuel and energy intensive. One kilogram of nitrogen fertiliser requires the energy equivalent of two litres of diesel. In 2000, energy used during fertiliser manufacture was equivalent to 191 billion litres of diesel, and is projected to rise to 277 billion litres in 2030. This is a major contributor to climate change, yet it is largely ignored. One kilogram of phosphate fertiliser requires half a litre of diesel.[20]

Nitrogen fertilisers also emit a greenhouse gas, nitrous oxide (N_2O), which is 300 times more destabilising for the climate system than CO_2.[21] According to the Intergovernmental Panel on Climate Change (IPCC), N_2O emissions have increased 45 percent due to an increased use of nitrogen fertilisers since the 1980s. Emissions over the life cycle of synthetic fertiliser manufacture and use are extremely significant. As per CIEL reports, according to the International Energy Agency (IEA), petroleum feedstock use for the fertiliser industry is projected to be 100 billion cubic meters in 2025. Since the 1950s, when fertilisers were pushed into the countries of the South through the Green Revolution, fertiliser use has increased ninefold, reaching a staggering 123 million tonnes in 2020.[22]

Synthetic nitrogen fertiliser is responsible for about 2 percent of global GHG emissions, which is more than the commercial

---

[19] Jake Marquez and Maren Morgan, 'The Legacy of the Men Who "Pulled Bread from Air"—A Reading by Maren', Death in the Garden, November 3, 2022, https://deathinthegarden.substack.com/p/45-the-legacy-of-the-men-who-pulled-beb.

[20] Vandana Shiva, *Soil Not Oil* (New Delhi: Women Unlimited, 2009).

[21] P.R. Shukla et al., eds., *Climate Change and Land*, Intergovernmental Panel on Climate Change (IPCC), 2019, https://www.ipcc.ch/srccl.

aviation sector, with 38 percent being a result of its production. The global life-cycle of CHG emissions from nitrogen fertiliser, as assessed by CIEL in their 2022 report, *Fossils, Fertilizers, and False Solutions*, is as follows:
- Transport: 29.8 million tonnes of CO_2
- Manufacturing: 438.5 million tonnes of CO_2
- CO_2 emissions from decomposition: 86.0 million tonnes
- N_2O from fertiliser applications: 379.9 million tonnes of CO_2e [carbon dioxide equivalent]
- Indirect N_2O emissions from nitrogen lost to waterways: 130.1 million tonnes of CO_2e
- Total emissions: 1,129.1 million tonnes of CO_2e

Since synthetic fertilisers are fossil fuel-based, they not only contribute to the disruption of the carbon cycle, but also disrupt the nitrogen cycle and the hydrologic cycle—both because chemical agriculture needs ten times more water to produce the same amount of food than organic farming and because it pollutes the water in rivers and oceans. Fertilisers rob the soil of nutrients and leave it desertified.

When the full costs of production, use and pollution from synthetic fertilisers are added, ecological, regenerative agriculture becomes an imperative. The alternatives to fertilisers that regenerate the soil are diverse. Pulses fix nitrogen non-violently in the soil, instead of increasing dependence on synthetic fertilisers produced violently by heating fossil

---

[22] Dana Drugmand et al., *Fossils, Fertilizers, and False Solutions: How Laundering Fossil Fuels in Agrochemicals Puts the Climate and the Planet at Risk*, Center for International Environmental Law (CIEL), 2022, https://www.ciel.org/wp-content/uploads/2022/10/Fossils-Fertilizers-and-False-Solutions.pdf.

fuels to 500°C. Chickpea can fix up to 140 kilos of nitrogen per hectare, and pigeon-pea can fix up to 200 kilos of nitrogen per hectare, non-violently. We don't need lab-made foods for proteins. Our dals provide protein to us while also fixing nitrogen for the soil.

Returning organic matter to the soil builds up soil nitrogen. Studies carried out by Navdanya over a period of twenty years comparing the soils of chemical and organic farms in the Doon Valley, Uttarakhand, revealed that organic farming increases the nitrogen content of soil between 44–144 percent, depending on the crops grown. The comparison was drawn between farms that were practicing chemical and organic farming on a continuous basis for a period of five years.[23]

Since war expertise does not provide us with expertise about how plants work, how the soil works, how ecological processes work, the potential of biodiversity and organic farming was totally ignored by the militarised model of industrial agriculture.

Integrating pulses in organic agriculture is the only sustainable path to food and nutritional security. This is the integration of life and the intensification of ecological processes, not the integration of power and intensification of chemicals, capital and control. Pulses are truly the pulse of life for the soil, for people and for the planet. They give life to the soil by providing nitrogen, and this is how ancient cultures enriched their soils. Farming did not begin with the Green Revolution

---

[23] Indira Rathore et al., 'A Comparison on Soil Biological Health on Continuous Organic and Inorganic Farming', *Horticulture International Journal* 2, no. 5 (October 2018), https://medcraveonline.com/HIJ/a-comparison-on-soil-biological-health-on-continuous-organic-and-inorganic-farming.html.

and synthetic nitrogen fertilisers. Whether it is the diversity-based systems of India—Navdanya and Baranaja—or the Three Sisters, planted by the First Nations in North America,[24] or the ancient milpa system of Mexico, beans and pulses were vital to indigenous agroecological systems.

Sir Albert Howard writes,

> Mixed crops are the rule. In this respect the cultivators of the Orient have followed Nature's method as seen in the primeval forest. Mixed cropping is perhaps most universal when the cereal crop is the main constituent. Crops like millets, wheat, barley, and maize are mixed with an appropriate subsidiary pulse, sometimes a species that ripens much later than the cereal. The pigeon-pea (*Cajanus indicus* Spreng.), perhaps the most important leguminous crop of the Gangetic alluvium, is grown either with millets or with maize...
>
> Leguminous plants are common. Although it was not till 1888, after a protracted controversy lasting thirty years, that Western science finally accepted as proved the important part played by pulse crops in enriching the soil, centuries of experience had taught the peasants of the East the same lesson.[25]

Vegetable protein from pulses is also at the heart of a balanced, nutritious diet for humans. The benevolent bean is central to the Mediterranean diet. India's food culture is based on 'dal roti' and 'dal chawal'. Urad, moong, masoor, chana, rajma, tur, lobia and gahat have been our staples. India was the largest producer of pulses in the world. Pulses have been displaced by

---

[24] 'Three Sisters' refers to the polyculture system practiced by Native Americans in North America in which they grow a combination of corn, beans and squash.

[25] Sir Albert Howard, *An Agricultural Testament* (London: Oxford University Press, 1921), 13, 14–15.

the Green Revolution monoculture, and now by the spread of monocultures of Bt Cotton and soyabean.

Biodiversity is a complex food web, teaming with earthworms, bacteria and fungi. Microfungi can fill between one-tenth and three-tenths of the area around plant roots, especially in undisturbed soils. Around 100 to 500 earthworms can live beneath 0.8 square meter (one square yard) of grassland or forest. Nearly 10 million nematodes, or unsegmented worms, can live beneath almost 0.8 square meter of soil. A 0.8 square meter patch of soil can contain 500 to 200,000 arthropods (insects such as beetles, ants and wood lice), spiders and mites. A teaspoon of healthy soil can contain between 100 million and 1 billion bacteria.[26]

It is this amazing biodiversity that maintains and rejuvenates soil fertility and supports agriculture. Living soil was forgotten for an entire century with very high costs to nature and society. An exaggerated claim was made that artificial fertilisers would increase food production and overcome all the ecological limits that land places on agriculture. Today, evidence is growing that artificial fertilisers have reduced soil fertility and food production and contributed to desertification, water scarcity and climate change. They have created dead zones in oceans.

Synthetic fertlisers destroy the soil-food web, thus undermining soil fertility and productivity. Moreover, their runoffs (90 percent) create dead zones and they emit N_2O, which is three times more damaging than CO_2. Fertiliser response, too,

---

[26] Catherine Arnold, 'Soil (and Its Inhabitants) by the Numbers', *Science News Explores*, February 25, 2021, https://www.snexplores.org/article/soil-and-its-inhabitants-by-the-numbers.

has reduced dramatically: over a period of thirty years, it went down from 13.4 kg grain/kg nutrient in 1970 to 3.7 kg grain/kg nutrient in 2005 in irrigated areas,[27] while in 1970, only 54 kg NPK/ha was required to produce around 2 t/ha, and some 218 kg NPK/ha was used in 2005 to sustain the same yield.[28]

Liebig was the first scientist to explain the role of nitrogen in plants, which was quickly appropriated by greed for commerce. A new industry was created for external inputs of nitrogen, dubbed 'growth stimulants'. Outraged at the distortion of his scientific findings, he wrote *The Search for Agricultural Recycling* in 1861. Liebig's book was the voice of a true scientist, protecting his truth from distortions of a pseudo-science being created by commercial interests. He writes,

> I thought it would be enough to just announce and spread the truth, as is customary in science. I finally came to understand that this wasn't right, and the altars of lies must be destroyed if we wish to give truth a fair chance.

The truth that Liebig was defending is that soil is alive, and its life depends on recycling. The lie he wanted to destroy is what he called the 'chemical hocus-pocus', that you can keep extracting nutrients from the soil, giving nothing back, and still obtain high yields. He questioned the false metric of 'yield', which merely measures the weight of the nutritionally empty

---

[27] For more information, refer to *Low and Declining Crop Response to Fertilizers*, Policy Paper no. 35, National Academy of Agricultural Sciences (New Delhi: NAAS, 2006), 8.

[28] P.P. Biswas and P.D. Sharma, 'A New Approach for Estimating Fertilizer Response Ratio: The Indian Scenario', *Indian J. Fert.* 4, no. 7 (2008): 59–62.

Industrial Farming and the Illusion of Food Security

commodity that leaves the farm. It is a measure of farming as an extractive industry, not of real farming. It is an illusionary measure that does not count the *total* biodiverse output, the *total* cost of inputs or the state of the soil and the farm. 'Yield' as a construct to promote fake farming based on chemical fertilisers artificially projects the *reduction* of nutrition per acre as an *increase* in food production. Liebig emphasises that what matters is care of the land, not 'yield of harvest', as well as the condition in which the field is left.

In total denial of climate science and soil ecology, Bill Gates is continuing the 'chemical hocus-pocus' when he says that we need to use more fertiliser. 'To grow crops, you want tons of nitrogen—way more than you would ever find in a natural setting. Adding nitrogen is how you get corn to grow 19 ft. tall and produce enormous quantities of seed.'[29] Selling more fertiliser is good for the profits of the chemical industry, but it is not good either for the soil or for the climate, or for our nutrition and health.

---

[29] Bill Gates, *How to Avoid a Climate Disaster: The Solutions We Have and the Breakthroughs We Need* (New York: Alfred A. Knopf, 2021), 123.

4 | Dead Food, Dead Metabolism

In the last few decades, our food, agriculture and health traditions have been ignored or destroyed under the assault of reductionist science and industrial systems of agriculture, combined with globalisation and free trade. The industrialisation and globalisation of food systems as well as the promotion of fast food and junk food are driven by chemical and industrial food corporations, leading to a climate crisis, an agrarian crisis, the erosion of biodiversity in agriculture, an increase in toxics in our food and a disease epidemic. Twenty percent of GHG emissions are accounted for by processed and packaged food and the food miles associated with globalised supply chains. Ultra-processed foods are energy intensive and resource intensive; they also use synthetic ingredients that are harmful to health, but have not been adequately tested.

The agrochemical industry and agribusiness, the junk food industry and the pharmaceutical industry profit while nature and nations get sicker and poorer. Powerful chemical corporations, controlling industrial agriculture and medicine, have tried to displace the ecological discourse of interrelatedness and interconnectedness with the reductionist paradigm they have shaped. Violent, toxic tools of industrial agriculture, industrial processing and industrial medicine have been elevated as human ends and been made the measure of human progress. Instead of humans assessing tools on the basis of their impact on the well-being of the planet and of people, they are used to assess the status of humanity. This has led to the imposition of a mechanistic paradigm on nature, on our minds and on our bodies.

Dead Food, Dead Metabolism

After impoverishing the biodiverse food basket and chemicalising our food (which then cannot be ingested properly, with the full range of nutrients and micronutrients, as our bodies are meant to ingest and digest systemically biological, not chemicalised, food), the agropharma industry comes forward with false solutions. We are offered supplementation which, as Dr Bhushan Patwardhan, distinguished professor of health sciences, biomedical scientist and ethnopharmacologist, says, 'has not demonstrated much preventive or therapeutic benefit'. He goes on to add, 'Rather than banking on exogenous antioxidant supplements, scientists are advising eating plenty of fruits, grains and vegetables, coupled with lifestyle modifications....'[1] The push to propose supplementation through genetic modification—of vitamin A rice or iron reinforced bananas—is fraught with non-results and the added risk of adverse consequences through the effects of genetic modification. Here again, biodiversity provides natural solutions.

It is not just our health that has suffered—it is also the health of the planet. Across India there is a water crisis because the so-called Green Revolution is extremely water intensive and has led to the diversion of rivers and the mining of groundwater for irrigation. More than 75 percent of our water resources have been destroyed and polluted due to chemical farming.

More than 90 percent of biodiversity in agriculture has disappeared on account of industrial agriculture, and with it, the diversity we need for a healthy and balanced diet. Chemical farming is based on cultivating monocultures of a few globally traded commodities.

---

[1] *Annam: Food as Health* (New Delhi: Navdanya, 2017), 133.

THE NATURE OF NATURE

As industrial agriculture and industrial food processing begin to dominate, epidemics of chronic, diet-related diseases start to spread. The Hopi describe this phenomenon as *powaqqatsi*—'an entity, a way of life, that consumes the life forces of beings in order to further its own life'. The powaqqatsi phenomenon is clearly in evidence today—we are dealing with a destructive force. And if corporations have their way, our fragile web of life will be poisoned beyond repair, the diversity of species will be driven to extinction and people will lose all freedoms to their seed, their food, their knowledge and their decision-making.

In spite of its rich scientific and intellectual heritage, based on food as health, India is rapidly emerging as the epicentre of chronic diseases, including cancer, obesity, diabetes and cardiovascular diseases, largely related to the kind of food produced and consumed. Health and disease begin in food, and food begins in agriculture and the soil. Traditional civilisations understood the intimate connection between agriculture, food, nutrition and health. Indigenous systems of science and agroecology evolved as sciences of life for healthy living and were anchored in propagating biodiverse, seasonal and local food.

We are at a watershed in terms of our agriculture, our food systems and our diets. For many decades now, the world has undergone a drastic change in the ways in which food is produced, processed and distributed: in agriculture, it has mostly gone from organic to chemical; in processing, from artisanal to industrial; and in distribution, from mainly local bazaars to supermarkets stocking food from around the world in defiance of seasonality.

Food, above all, is life-giving and nurturing—and yet, in our lives, we find that this primary function has been vitiated. How

Dead Food, Dead Metabolism

has this happened? To start with, it may be a platitude, but we become what we eat; and since, directly or indirectly, our food comes from plants, the health of the plant will translate into the health of our bodies. So, the plants which have ingested the poisons from so-called fertilisers transfer these to all those who feed upon them and, as we have seen, these can have drastic consequences, such as poor nutritional intake, impaired digestive systems and cancer, to name a few.

As food travels from the fields to our tables, it goes through two more stages: processing and retailing. Traditionally, processing was in women's hands and was artisanal for the most part. Women cleaned the wheat, washed it, dried it and then proceeded to grind it by hand for flour; till about twenty years ago, one could still go to a chakkiwala (local flour mill) and have one's wheat ground into flour. However, with the advent of the likes of Cargill and Pillsbury, shops now retail industrially-processed atta (flour) from which most of the nutrients have been stripped, having gone through high-intensity processing machines that heat up and thus destroy precious nutrients. This, of course, works to the advantage of the food industry, which can then get into the 'fortification' business.

Industrial processing uses high-intensity processing machines for grinding as well as for extruding and applying pressure, all of which lead to structural changes in the food with negative consequences on health. Most industrially-produced snacks are subjected to these denaturing processes. Moreover, many oils, e.g., soyabean oil and canola oil, are subjected to denaturing processes in order to extract the maximum volume.

If we look at foods like packaged cereal, snacks, confectionery and suchlike, their processing contains a cocktail of chemical additives in the form of: preservatives, to prevent them from rotting; colorants, to make food more attractive or 'natural'; taste enhancers and flavours, to increase palatability or to impart a certain flavour; texturants, to give a particular texture to the food; and stabilisers, to increase shelf life. To make matters worse, processed food manufacturers do not always list all the additives used on their labels, especially as consumers are now more savvy at decoding them.

The industrial meat industry is another danger zone where not only are animals kept in cruelly dismal and inhuman conditions, but are also fed heavy doses of hormones or antibiotics against disease and to increase yield, producing more dead and toxic food.

Before food reaches the supermarket shelves, it has to be packaged; very often the packaging consists of plastic bags or containers or aluminium foil, highly unecological as well as sources for further contamination of the food they contain, since the chemicals in plastic and aluminium always leach into the content. It is ironic that this food, contaminated in more ways than one with chemicals at the processing levels and chemicals from the packet/containers, is considered safer than what is sold in the open. Research by eminent scientists, such as the French molecular biologist Dr Gilles-Éric Séralini and others, points to the fact that food which has become less and less natural, not to say denatured altogether, interferes with our brain function when consumed. It can cause inflammation in our body, can become addictive and, in short, can lead to foodstyle diseases. Dead food results in dead metabolism.

Following globalisation and liberalisation, the market, everywhere in the world, has become a world market; throughout the year, goods from all over the planet adorn its shelves. As Marion Nestle points out in her book *Food Politics:*

Dead Food, Dead Metabolism

How the Food Industry Influences Nutrition and Health, a veritable food strategy is put into place by the food industry to entice people into consuming more and more, regardless of the maximum intake considered healthy, fuelling obesity which, in any case, is already imminent via processing and the use of high fructose corn syrup, trans fats, etc. After all, the excess food produced has to find takers to satisfy the high-volume model of production.

Then there is the expiry date battle. Given that liberalisation means countries that have produced in excess or are producing items no longer popular in their markets can dump their goods on markets in the South, there is a big expiry date scam in practice. Often, consumers who are not attentive to such details will buy food that is no longer fit to be consumed.

Another consequence of globalisation/liberalisation is that all food items, from mangoes to melons, which are actually seasonal, are available throughout the year. Our bodies are tuned to eat as per the seasons; but, additionally, there are socio-ecological effects, since local farmers are often marginalised. Unaware consumers seek the exotic over the local—items that have a very heavy ecological footprint, having accumulated large amounts of food miles.

The spread of junk food was a result of dismantling the regulatory frameworks that countries had evolved to protect the livelihoods and health of their citizens. The Sanitary and Phytosanitary (SPS) Agreement of the WTO, drafted by the junk food industry, forced open the markets of the world to the sale of ultra-processed food. They altered the science of food production and food processing, and they neutralised the regulatory framework. Sales of processed food increased, and with it, the non-communicable chronic disease pandemic.

In such a scenario, what is the consumer, the co-producer of agriculture, to do? Nestle says, '[Choose] food according to freshness, taste, nutrition and health, but also social and environmental issues.'

Food safety and foodstyle diseases

The issue of food safety is progressively looming larger and larger with the hyper industrialisation and globalisation of food. According to Tim Lang, Professor of Food Policy at City University, London,

> Incidence of food-borne disease has in fact risen during the era of the productionist paradigm. In West Germany cases of *S. Enterites* infections rose from 11 per 100,000 head of population in 1963 to 193 per 100,000 in 1999, while in England and Wales formal notifications of the same disease rose from 14,253 cases in 1987 to 86,528 in 2000.

For his part, Colin Tudge, a British biologist, observes,

> The modern food supply chain is convoluted and so long that it allows endless opportunities for malpractice of all kinds—including many that beggar the imagination of those who are not criminally inclined. . . Sometimes though, it is not at all easy to draw a line between outright villainy (like the adding of contaminants) from the standard, legitimate practices of the modern food industry.[2]

The early 2000s saw several epidemics arising from the way food is produced industrially: in the UK, the mad cow disease

---

[2] Vandana Shiva, 'A Law for Food Fascism: Part One', Slow Food, June 23, 2005, https://www.slowfood.com/blog-and-news/a-law-for-food-facism-part-one.

Dead Food, Dead Metabolism

jumped from cattle to humans. New strains of *E. coli* led to thousands of food poisoning cases in the US, causing the death of 5,000 people. The swine flu fever in Asia caused havoc, leading to millions of pigs being killed, with the new Nipah strain killing hundreds of pig-farm workers. The spectre of avian flu raises regularly its head.

New food safety laws are deregulating large corporations while overregulating small-scale, self-organised economies. We cannot impose the same laws for toxic and GM foods and for local, artisanal processing as well as small kiosks and outlets; the former need stricter laws than the latter, which are not a toxic threat or consist of centralised producers needing centralised regulation.

These new laws are the basis of the WTO's SPS Agreement. India's diverse, decentralised edible oil industry offers an example of how inappropriate standards were used to destroy this sector. August 1998 saw the introduction of a new packaging order for edible oils on grounds of food safety; its consequence was the shutting down of millions of small-scale local oil mills. This had a huge impact on local edible oils like mustard oil, which was, in fact, banned. Combined with the WTO trade rules that remove import restrictions, false food safety laws flooded India's markets with oil from genetically-engineered soyabean and palm oil.

India has a variety of oilseeds such as coconut, groundnut, linseed, mustard, sunflower and sesame; biodiversity going hand in hand with cultural diversity. The main consequence of the mustard oil ban and the ban on selling edible oils in unpackaged form was an adverse impact on our oilseed biodiversity and the diversity of our edible oils and food cultures. It was also a destruction of economic democracy and economic freedom to produce oils locally, according to locally available resources and locally appropriate food cultures.

Since indigenous oilseeds are high in oil content, they can be processed at household or community level, with eco-friendly, decentralised and democratic technologies. They therefore mean freedom for nature, our farmers, our diverse food cultures and the rights of poor consumers.

By contrast, soyabean oil signifies concentration of power and the colonisation of nature, cultures, farmers and consumers from the seed to trade, processing and packaging. Because of its low oil content, the extraction of soyabean oil needs heavy processing which is environmentally unfriendly and unsafe for health.

Monsanto controls seeds through its patents and ownership of seed corporations. Cargill, Continental AG, and other trading giants control trade and milling operations internationally. The manipulation of oil prices and the restrictions placed on indigenous oilseed processing and sales forced Indians to consume soyabean and palm oils, further strengthening a monoculture and monopoly system.

Free trade and economic globalisation have been projected as economic freedom for all. However, as the case of the mustard oil crisis and soyabean and palm oil import reveals, so-called free trade is based on many levels of the destruction of small producers, processors and poor consumers.

The application of food safety regulation clearly demonstrates how unequally small and big players are treated. The government of India was prompt in immediately banning mustard oil but did nothing to prevent the dumping of toxic, genetically-engineered soyabean. Be it India, the US or countries around the world, governments do not penalise global players for the toxicity of their food. The simple act of buying edible oils, or even other

Dead Food, Dead Metabolism

food, reveals the degree of unfairness and lack of democracy as far as food production is concerned.

The 'choice' of junk food by societies was in a way facilitated by the deregulation of authentic food safety laws and their replacement by corporate-written laws favouring industrial food production, as well as policing and criminalising local, indigenous, artisanal production and processing.

Europe is not exempt from food safety laws that are heavily tilted towards big corporations. There, too, such laws were threatening small producers of typical foods. To protest this, the Slow Food Movement organised half a million signatures that forced the Italian government, for example, to amend a law that would have compelled even the smallest food producer to conform to the pseudo hygienic standards that suit corporations like Kraft.

Different foods have different risks and need different safety laws and systems of management. This is why Europe has different standards for organic, industrial and genetically-engineered foods. Organic standards are set by organic movements, while standards for genetically-engineered foods are set at the European level through the Novel Food laws. In addition, there is the movement to protect the cultural diversity of food through the classification of 'unique' and 'typical' foods. These standards are cultural, based on indigenous science and community control, not on industrial 'science' controlled by governments manipulated by food giants like Cargill, ConAgra, Lever, Nestlé or Phillip Morris, as well as gene giants like Monsanto. Carving out these spaces of freedom has been the contribution of the Slow Food Movement.

Ideally, India, like Europe, should have had three different levels of regulation for three different systems of food production: one at the local and community level for small-scale processing; a second for industrial processing, geared to the prevention of food adulteration, updated periodically to prevent new food hazards; and a third, a GMO food law, controlling imports, labelling, segregation, traceability and so on.

We, however, now have the 2005 Food Safety and Standards Authority of India (FSSAI) which caters to all three categories. This kind of centralised law represents the worst form of license and inspector raj and operates like food fascism, based on a food mafia serving global corporations. Our food freedom, livelihoods, our food safety and diversity are all at stake with the FSSAI. It can be used to criminalise every small roadside restaurant owner and street vendor who is not introducing obesity or diabetes, cancer or heart disease; they provide affordable dal and roti to millions of working people.

How food is processed determines its quality, nutrition and safety. Home-processed bread is not the same as mass-produced industrial bread. They are not 'like products', in WTO jargon; they are quite different in terms of their ecological content and public health impact. A factory chicken is not the same as a free-range chicken, both in terms of animal welfare and in terms of food quality and safety. GMO corn is not the same as organic corn. The former contains antibiotic resistance markers, viruses used as promoters and genes for producing toxins such as Bt. Regulating Bt corn for safety needs a different system than that for organic corn, just as factory farming needs different regulatory processes from organic farming.

A pluralism of production processes and products needs a pluralism of laws, and science that is appropriate for the safety issues and governance systems that a product or production process demands. Chemical processing needs chemistry labs and

Dead Food, Dead Metabolism

chemists; GMOs need genetic ID laws; and organic processing needs indigenous science and community control. But in the case of GMOs, there are no international standards. There are European laws on Novel Foods and an absolute deregulation of GM foods in the US.

On May 13, 2003, the US, together with Canada and Argentina, challenged Europe's moratorium on GM crops and foods. Arguing that their GM products were being unfairly discriminated against, they challenged the precautionary principle in decision-making about GM crops that is embedded in European decision-making.

The EU regulations take into account the EU's international trade commitments as well as the requirements of the Cartagena Protocol on Biosafety, with respect to the obligations of importers. The EU's regulatory system for GMO authorisation is in line with WTO rules: it is an attack on the use of the precautionary principle with clear, transparent and non-discriminatory information. There is therefore no issue that the WTO needs to examine.

Many countries are now looking at EU regulations to develop their own policies. The US fears that several countries will adopt a similar approach to regulate GMOs and GM food and feed products. The Swiss GM legislation that came into force on January 1, 2004, is a good example.

The Swiss law is more stringent than current EU legislation on liability and co-existence aspects, and is based on the precautionary and 'the polluter pays' principles (Article 1). It aims to protect the health and security of human beings, animals and the environment. It also aims to permanently maintain biological diversity and soil fertility, and to allow freedom of choice to consumers. The EU Moratorium represents the will of its people not to be force-fed. It crystallises (as the patent on seeds still does) the worldwide mobilisation of people against the reinterpretation of national security and sovereignty to facilitate the global control by US corporations over resources and markets.

If Europe had not suspended its approvals process in 1998, there would have been some very undesirable consequences, as the indirect effects of growing GM herbicide-tolerant (HT) crops on farmland wildlife would not have been taken into account. There would not have been a requirement for monitoring environmental or human health, consumers would not have been able to make a choice regarding products derived from GM crops and no labelling and traceability requirement would have been introduced.

Food in India is processed variously for consumption. There is the first level of artisanal processing, which has been in existence for many years. Wheat was processed into flour in stone grinders, first operated manually and then mechanically, with minimum power. So, too, with pulses, and oilseeds were similarly cold pressed into oil. As far as snacks and pickles were concerned, they were mostly processed as a home industry under the supervision of women. These activities still continue at various levels: whether for a niche community of aware consumers or for those who can only afford home processing.

At the same time, there is the industrial-scale food processing of wheat and other grains, the polishing of rice, the production of white refined sugar, refined oils and so on. This mode of food processing relies heavily on fossil fuels, chemicals and packaging material made of plastic or aluminium foil.

In 2005, the Indian government drafted a Food Safety and Standards Bill as an 'integrated food law', prepared with the intention of being contemporary and comprehensive, ensuring

Dead Food, Dead Metabolism

better consumer safety through food safety management systems, while setting standards based on science and transparency and also meeting the dynamic requirements of international trade and Indian food, trade and industry. It soon transpired, however, that the law actually helps to lubricate international trade and the expansion of global agribusiness. In a way, the 2005 Food Safety law legalises the adulteration of our entire food system with toxic chemicals, various additives, food colours, stabilisers and taste enhancers for processed and ultra-processed foods. It reinforces a centralised mode of production over decentralised, more ecological ones, and ignores the most distinctive aspects of India's food systems: indigenous science, cultural diversity and local economic livelihoods, even though a sizeable percentage of our food is processed naturally and locally for local consumption and sale.

In the early 2000s, a case was filed against Coca-Cola and PepsiCo regarding the presence of high levels of pesticides in their beverages, which led to the creation of a Joint Parliamentary Committee to look into the matter. Their report, as well as other studies, clearly indicated that both soft drink manufacturers were using significant quantities of very harmful chemicals in their drinks to make them more addictive, in blatant disregard of public health. These chemicals included large amounts of phosphoric acid, caffeine high fructose corn syrup, all known to have increased the prevalence of Non-Alcoholic Fatty Liver Disease (NAFLD), as well as sugar, ethylene glycol and carbon dioxide. Yet, despite an order by the Rajasthan High Court to disclose the contents of their beverages, both companies refused on the pretext of safeguarding trade secrets. Earlier Prevention of Food Adulteration Acts (PFAs) required companies to disclose the ingredients of all packaged food on their labels. However, in

2004 the Supreme Court stayed the Rajasthan High Court order and Coca-Cola and PepsiCo were let off the hook.

The results of the deregulation of food safety are there for all to see. Instead of being regulated, corporations like Coca-Cola, Nestlé, Kellogg's and the Gates Foundation are now sponsoring the Annual Conference of the Nutrition Society of India.[3]

In India's diverse, decentralised plural economy, a centralised, integrated law is inappropriate on many counts. One law for all food systems is a law that privileges large-scale industrial commercial establishments while discriminating against and criminalising the small, the local, the diverse. It places kitchens and kiosks, cottage and household industries in the same category as the Nestlés, Cargills and ConAgras. Domestic and local consumption, including not-for-profit food provisioning, should not be on par with the import of hazardous GMOs. This unscientific system serves the purpose of disease-spreading, and of unecological MNCs, at the cost of decentralised, plural food systems that are both healthy and ecological, while at the same time destroying millions of livelihoods and millennia of diverse gastronomic traditions.

The Food Safety and Standards Act, 2006 creates a culture of 'food fascism'. Article 16 (6) of the Functions of Food Authority states: 'The Food Authority shall not disclose or cause to be disclosed to third parties confidential information that it receives for which confidential treatment has been requested and has been acceded.'

Risk assessment in the hands of centralised, corruptible agencies offers no protection for consumers, as the disease and

---

[3] Arun Gupta (@Moveribfan), 'Terrible news. . .,' Twitter, November 25, 2023, 3:59 a.m., https://twitter.com/moveribfan/status/1728337636874764345?s=58.

Dead Food, Dead Metabolism

health epidemic in the US linked to over-processed, industrial foods has shown. Even while the US is at the epicentre of food-related public health crises, the US government works to export its food laws elsewhere, to deregulate the food industry and over-regulate ordinary citizens and small enterprises.

Industrial foods need chemical labs, genetically-engineered foods need genetic ID labs, but cooking fresh dal and roti does not need testing for toxic chemicals and transgenes. The risk and safety standards for lassi (sweet yoghurt drink) in a roadside eatery and a synthetic milkshake at a fast-food chain must be treated differently. In his bestseller, *Fast Food Nation*, Eric Schlosser presents a list of the deadly cocktail contained in the artificial flavour of Burger King's strawberry milkshake, which includes amyl acetate, amyl butyrate, amyl valerate, anethole, anisyl formate, benzyl acetate, benzyl isobutyrate, butyric acid—the list goes on.

Pseudo safety laws prevent any act of private community feeding, such as at church gatherings, for example. In India, there are plural community feeding practices—langar in gurudwaras, zakat in mosques or bhandaras during Hindu prayers—through which thousands of people are fed. Additionally, the livelihoods of chaiwalas and dhabawalas, as well as the entire household and cottage industry, are rendered illegal in food processing, leading to criminalising safe foods while legalising the food crimes of disease-producing agribusinesses.

Pseudo safety laws encourage poverty and unemployment by destroying millions of small-scale livelihoods in food production and processing. A just and equitable food law would recognise that decentralised food economies enhance nutrition, safety, culture and livelihoods. The FSSAI, on the other hand, prejudices indigenous food systems in various ways, one of which has to do with the labelling of GMO foods. It states: 'All food products having total Genetically-Engineered (GE) ingredients [of] 5% or more shall be labelled.' In other words, up to 5 percent need not be labelled. This is misleading and unfair as well as in defiance of consumer rights. Dominant international standards do not allow the presence of more than 1 percent of GMOs.

There are many examples in the rules for labelling that are unscientific, misleading and in effect a means of imposing unhealthy industrial foods on Indian citizens.

Example 1: Gluten
The proposed rules state that all flour will include a label on gluten. This is irrelevant, as in India there are several local wheat varieties that have very low gluten content, so much so that Monsanto tried to take out a patent (No. 0445929B 1) on the naphal variety. (Navdanya challenged this biopiracy patent, along with Greenpeace and Bharat Krishak Samaj, at the European patent office on January 27, 2004, and won.)[4]

It is the industrial production of wheat, based on uniformity, combined with industrial processing, which damages its structure and has led to an epidemic of gluten allergies.

Example 2: Fake fibres
Industrial flour, called 'refined' and 'enriched', is stripped of its nutrients and fibre. Synthetic nutrients and fibres are then added to processed flour. The FSSAI rules have a section on fake fibres which is detrimental to the fibre-rich whole flours of diverse grains in our diets.

'Dietary fibre' now includes

---

[4] For more, see Vandana Shiva, *Origin: The Corporate War on Nature and Culture* (Delhi: Natraj Books, 2018).

- *Carbohydrate polymers, which have been obtained from food raw material by physical, enzymatic or chemical means.*
- *Synthetic carbohydrate polymers.*

We need process labelling that clearly describes the process of production and processing, not an unscientific product labelling based on equating unhealthy foods and healthy foods and concealing the health hazards of fake food.

Example 3: Healthy fats and trans fats

India, the land of rich biodiversity in edible oils—mustard, sesame, coconut, linseed, groundnut, etc.—is now 70 percent dependent on imports of palm oil and GM soyabean oil. Solvent extraction plants use a neurotoxin, hexane, to extract the oil from soyabean or palm.

The FSSAI equates bad fats which are unfit for eating with good fats necessary for health: '"Fat" means total lipids including saturated fat, monounsaturated fat, polyunsaturated fat and trans fat.'

Yet, the world over, trans fats are being removed from food.

Example 4: Sugars

The FSSAI unscientifically defines sugars in terms of chemical reductionism, equating good and bad sugar instead of assessing the quality and method of processing: '"Sugars" means all monosaccharides (glucose, fructose, etc.) and disaccharides (maltose, sucrose, lactose, etc.).'

However, gur, or unrefined sugar, is very different in the way it is processed (and in quality and health impact) from industrial sugar and fake sugars such as high fructose corn syrup.

According to the World Health Organization, overall, the share of ultra-processed foods in the retail sales of total processed food (essential/staple plus ultra-processed) was 43 percent in 2011 and 36 percent in 2021. It is projected to be 39 percent by 2032.[5] It is estimated that consuming these foods encourages people to eat more, consuming ~500 additional calories per day and gaining a significant amount of weight, ~900 grams on average in two weeks.[6] Studies found that a 10 percent increase in the consumption of ultra-processed foods can increase the risk of diabetes by 15 percent and lead to higher premature mortality due to cardiovascular diseases.

Most of the chronic diseases of our times are foodstyle, rather than lifestyle, diseases; food-related diseases such as neurological problems, metabolic disorders like obesity and diabetes, cancer, infertility and intestinal problems are on the rise. They are called foodstyle diseases because they are associated with changing diets and the foods we are increasingly consuming—chemically grown and highly processed. How we grow our food, how we process it and what we eat are major ecological determinants of health and disease. Industrially-processed food and a homogeneous, uniform diet lacking in diversity is unable to provide the nourishment that the trillions of cells in our bodies need to perform the diversity of functions that make us mentally and physically healthy.

---

[5] *The Growth of Ultra-Processed Foods in India: An Analysis of Trends, Issues and Policy Recommendations* (New Delhi: World Health Organization, Country Office for India, 2023), https://www.who.int/publications/i/item/9789290210672.

[6] Kevin D. Hall et al., 'Ultra-Processed Diets Cause Excess Calorie Intake and Weight Gain: An Inpatient Randomized Controlled Trial of *Ad Libitum* Food Intake', *Cell Metabolism* 30, no. 1 (July 2019), https://www.cell.com/cell-metabolism/fulltext/S1550-4131(19)30248-7.

Dead Food, Dead Metabolism

According to the National Institute of Health,

> Heart disease is now the world's leading cause of death, claiming 17.3 million lives each year. . . . Latest statistics suggest that in India, there are roughly 30 million heart patients. . .[of which] 14 million reside in urban areas and 16 million in rural areas. . . and two lakh surgeries are being performed every year. . . . India has seen a rapid transition in its heart disease burden over the past couple of decades. . . [and] atherothrombotic cardiovascular diseases in India will surpass any other country in the world.[7]

Diabetes is fast gaining the status of a potential epidemic in India with more than 62 million diabetic individuals currently diagnosed with the disease. In 2000, India topped the world with the highest number of people with diabetes mellitus (31.7 million), followed by China (20.8 million) and the United States (17.7 million) in second and third place, respectively. The prevalence of diabetes is predicted to double globally from 171 million in 2000 to 366 million in 2030, with a maximum increase in India. It is estimated that by 2030, diabetes mellitus may afflict up to 79.4 million individuals in India, while China (42.3 million) and the United States (30.3 million) will also see significant increases in those affected.[8]

Cancer, also a major disease in the country, is affecting rural and urban India alike. The government-run National Cancer Control Programme estimates that there are between 2 and

---

[7] 'World Heart Day 2015: Heart Disease in India Is a Growing Concern, Ansari', Press Trust of India, July 13, 2017, https://food.ndtv.com/health/world-heart-day-2015-heart-disease-in-india-is-a-growing-concern-ansari-1224160.

[8] Seema Abhijeet Kaveeshwar and Jon Cornwall, 'The Current State of Diabetes Mellitus in India', *Australas Med J* 7, no. 1 (January 2014), https://www.ncbi.nlm.nih.gov/pmc/articles/PMC3920109.

2.5 million cancer patients in the country at any given point of time. Recent reports suggest that India is seeing one million new cases of cancer every year, and about 1,300 people die each day because of it. Doctors fear that South Asia (India, Pakistan and Bangladesh) could account for nearly 50 percent of the world's cancer cases by 2040.[9] In 2010, 556,400 people across the country died of various types of cancer, according to one study, and the 30–69 cohort, the productive age group, accounted for 71 percent or 395,400 of the deaths.[10]

Food grown with chemicals affects our health in three ways. First, since it is focussed on a handful of commodities, most of which go towards biofuel and animal feed, it contributes to hunger and malnutrition. Ninety percent of the corn and soyabean produced in the world is used as biofuel and animal feed, not for feeding humans. Only 30 percent of the food we eat comes from large industrial farms; 70 percent is derived from small farms, which use only 20 percent of available agricultural land. Second, since industrial agriculture produces uniform, homogeneous monocultures, it contributes to diseases linked to deficiencies of vital, diverse nutrients. Third, toxics used in agriculture make their way into our food and are a factor in diseases such as cancer.

Herbicides like glyphosate, sold under the brand name Roundup, have been recognised by the WHO as a probable

---

[9] Sumitra Debroy, 'Cancer Cases in India Estimated to Rise from 14 Lakh Per Year to 20 Lakh by 2040', *The Times of India*, November 5, 2023, https://timesofindia.indiatimes.com/india/cancer-cases-in-india-estimated-to-rise-from-14-lakh-per-year-to-20-lakh-by-2040/articleshow/104977239.cms.

[10] 'The Cruel Economics of Cancer Care', Money Control, September 29, 2015, https://www.moneycontrol.com/news/trends/ge/the-cruel-economicscancer-care-1487261.html.

Dead Food, Dead Metabolism

carcinogen; the US state of California is demanding that it be labelled as such. Monsanto's attempt to block the labelling order was legally defeated. Many regions in Europe are banning its use. Scientists have found Roundup implicated in kidney failure and liver diseases. Sri Lanka has banned Roundup because it was linked to kidney failure.

Only two applications of GMOs have been commercialised—Roundup Ready crops which are supposed to control weeds, and Bt toxin crops which are supposed to control pests. GMOs are associated with major food safety and biosafety issues; for example, the proposed GMO mustard uses genes for terminating fertility, genes for herbicide resistance and viral genes as promoters. No animal feeding studies have been conducted yet; the partial tests that have been done are with surrogate proteins. The unscientific assessment of the safety of GMOs is based on a false premise: 'substantial equivalence', which assumes that non-GMO food is equivalent to GMO food, thus creating a fallacy for asserting safety. However, as mad cow disease (BSE, bovine spongiform encephalopathy) showed, while food may be substantially equivalent on the basis of 'nutritionism', if it is *structurally inequivalent*, the consequences can be serious. Mad cow disease was identified as having been caused by 'prions'—deformed proteins—which became self-infecting agents that spread BSE.[11]

High fructose corn syrup (HFSC), increasingly being used as a sweetener in industrial soft drinks and by Coca-Cola and PepsiCo, has been identified as being responsible

---

[11] Bovine Spongiform Encephalopathy (BSE), or mad cow disease, US Centers for Disease Control and Prevention, https://www.cdc.gov/mad-cow/php/animal-health/?CDC_AAref_Val=https://www.cdc.gov/prions/bse/index.html.

for the rise of Non-Alcoholic Fatty Liver Disease. A study in the journal *Metabolism*[12] found that in the 35 years following the introduction of HFCS, the rise in obesity has paralleled an increase in the use of HFCS. It has led to skyrocketing insulin production, while suppressing the response to leptin, which regulates appetite. With the disruption of its regulatory mechanisms, the body begins to store fat and obesity is the result.

Soft drink companies also use colour additives. The artificial caramel colouring in colas is made by heating ammonia and sulphites under high temperatures which produces a cancerous substance called 4 methylimidazole (4-MEI). In 2007, a US government study concluded that 4-MEI caused cancer in mice; and in 2011, a study by the International Agency for Research on Cancer determined that the chemical is a probable carcinogen.

Artisanal coconut and mustard oils are now being recognised as healthy, despite all the pseudo-scientific propaganda against them for decades by the industrial food processing lobby; it has been promoting trans fats by influencing food policy, trade policy and scientific research, and by spending enormous amounts of money on misleading advertisements. Trans fats help increase the shelf life of processed food, allowing it to remain solid at room temperature. According to a 2012 study published in the *Annals of Internal Medicine*, a mere 40 calorie per day increase in trans fats in your diet increases the risks of heart disease by 23 percent. The Centers for Disease Control has also attributed

---

[12] MyPhuong T. Le et al., 'Effects of High-Fructose Corn Syrup and Sucrose on the Pharmacokinetics of Fructose and Acute Metabolic and Hemodynamic Responses in Healthy Subjects', *Metabolism: Clinical and Experimental* 61, no. 5 (May 2012): 641–51, https://doi.org/10.1016/j.metabol.2011.09.013.

heart attacks to trans fats. It is worth noting that trans fats were originally invented to make candles; Procter & Gamble took out a patent on them and used them to produce cheap food.

The artisanal versus industrialised processing of wheat, salt and lentils, to name a few, offer similar examples of the takeover of small enterprises and localised production by agribusinesses and multinational corporations. The same handful of corporations sell both toxic agrichemicals as inputs in industrial agriculture and pharmaceutical products as inputs in industrial medicine. Disease is a market opportunity for this agro-medical-industrial complex. It is not in their interest to promote internal-input agriculture or health systems based on the science of the body as a living, self-regulating system; nor is it in their interest to understand the links between agriculture and health. Profits are ensured by selling new 'magic bullets' which will not cure and may, instead, create new complications.

Ultra-processed food contributes to emissions that destabilise the climate and disrupt the earth's metabolism. Industrially-produced food is at the root of both the climate crisis and the health crisis. Going further and faster down the resource intensive, energy intensive and disease-inducing path will aggravate both the climate and health catastrophes.

There is an intimate connection between the biodiversity in the soil, the biodiversity of our plants and the biodiversity in our gut, between ecological sustainability and health.

Junk energy, junk food

My quantum worldview and ecological perspective have taught me that the principles of non-separability and potential are central to both the quantum and the ecological paradigms. Over the years I have seen the health of ecosystems and organisms as a non-localised symphony in quantum coherence. In this perspective, ecological degradation and disease impair the self-organisation, self-regulation, self-healing and renewal capacity of living systems. In an ecological paradigm, agriculture, food production and health are internal-inputs systems, which have the capacity and potential to produce what they need. The earth, food and our bodies are interconnected living systems; the health of the planet and our health are a continuum.

We are one humanity on one planet, united through biological and cultural diversity. Issues of sustainability and justice are interconnected. Processes that harm the earth harm humans. A healthy planet, healthy ecosystems, healthy organisms and healthy people are all based on diversity, balance, symbiosis. When we respect the integrity and limits of the earth, her ecosystems and species, we are connected through health; when we violate the integrity of living beings and the ecological limits of the earth, we are connected through disease. And when we destroy the biosphere and substitute living processes with fossil energy and fossil industrialisation, we destabilise the earth's self-regulating mechanism that stabilises climate systems.

Food is the currency of life. It is the connection between us, the earth and other species. We are part of the web of life, which is a food web. The food web weaves the web of life in cooperation and mutuality. Food is the metabolism that connects humans with the earth, with biodiversity and cultural diversity.

When we destroy biodiversity within us, we create chronic diseases. When we destroy the biodiversity of our gut microbiome with industrial food and ultra-processed food, chronic diseases like diabetes and cancer take on epidemic proportions. The biodiversity outside and the biodiversity within

Dead Food, Dead Metabolism

us in our gut microbiome are interrelated; destroying one, we destroy the other.

The greater the biodiversity in any ecosystem, the greater its resilience and resistance to disease. This also applies to our gut ecosystem. Biodiversity destruction in the gut microbiome is responsible for inflammation and metabolic dysregulation, leading to many chronic diseases. When our gut biodiversity plummets because of the toxics or deficiencies in the food we eat, health pandemics can emerge—gastrointestinal infections, autoimmune diseases like asthma, rheumatoid arthritis, inflammatory bowel disease, autism spectrum disorders, obesity and metabolic diseases.

When it comes to exposing the damage that an industrial system does to the planet's biodiversity, it is necessary to understand the extent to which human beings are part of that same biodiversity and how greatly they share the risks. It is no coincidence that more and more researchers are focusing on the relationship between the loss of biodiversity and an increase in inflammatory diseases. The decline in our immune system's ability to properly function is associated with the state of health of our microbiome, which is the system of bacteria, viruses, fungi, yeasts and protozoa in our intestines. Also called our 'second brain' by scientists, it performs a number of important functions that significantly contribute to the health of our immune system. A poor functioning microbiome, or its lack of diversity, also entails greater risks of developing various neuropsychiatric disorders such as depression, schizophrenia, autism and anxiety. Additionally, recent research has confirmed that the composition and diversity of the microbiome are very important in determining anti-tumour immunity. The fact that the human microbiome is in distress is confirmed by a practice that is becoming widespread in medical circles: namely, the transplantation of faeces, aimed at transferring a healthy microbiome into a patient whose microbiome is no longer functional.

As Salvatore Ceccarelli, international expert in agronomy and plant genetics, explains:

> How can we have a diet based on diversity, if six percent of our calories come from just three plant species: wheat, rice and corn? And how can we have a diet based on diversity, if almost all the food we eat is produced from seed varieties that, in order to be legally traded, must be registered in a catalogue that is called a register of varieties; and that, in order to be recorded in this register, must be uniform, stable and recognisable? Between the need to eat 'diverse' foods discussed so far, and the uniformity in food products required by laws on crops, there is a clear contradiction.[13]

Fast food and rising metabolic syndrome

Industrial agriculture not only limits the varieties of food — it also places large quantities of food with very low nutritional value in the market.

Nadia El-Hage, former FAO scientist, notes,

> The food that is being put on the market today is not of the same quality as before the Second World War; compared to more than

---

[13] Comment at the meeting of a group of experts in Florence, Italy, in 2018 to draft the *Manifesto on Food for Health*. International Commission on the Future of Food and Agriculture, *Manifesto on Food for Health: Cultivating Biodiversity, Cultivating Health* (New Delhi: Navdanya / RFSTE, 2019), https://navdanyainternational.org/wp-content/uploads/2019/05/Manifesto-Food-for-Health-_-1-5-2019.pdf.

60 years ago, most crops have lost, on average, almost 20% of nutrients with peaks of up to 70 or 90%.[14]

Junk food and ultra-processed food are contributing to an emerging epidemic of non-communicable chronic diseases; these are now being considered as metabolic disorders and are collectively referred to as the metabolic syndrome (MetS). MetS is a collection of diseases related to metabolic abnormalities, including abdominal obesity, hypertriglyceridemia, low high-density lipoprotein-cholesterol (HDL-C) concentrations, hypertension and hyperglycaemia (which have a strong association with the development of type 2 diabetes), and cardiovascular morbidity and mortality in adults. The prevalence and incidence of MetS is increasing rapidly in

---

[14] Manlio Masucci, 'Food for Health: The Right to Health Is to Live Healthy Lives,' Navdanya International, March 6, 2020, https://navdanyainternational.org/food-for-health-the-right-to-health-is-to-live-healthy-lives.

[15] Golaleh Asghari et al., 'Fast Food Intake Increases the Incidence of Metabolic Syndrome in Children and Adolescents: Tehran Lipid and Glucose Study', *PLoS One* 10, no. 10 (October 2015): e0139641, https://www.ncbi.nlm.nih.gov/pmc/articles/PMC4598125. See also Letícia Ferreira Tavares et al., 'Relationship between Ultra-Processed Foods and Metabolic Syndrome in Adolescents from a Brazilian Family Doctor Program', *Public Health Nutrition* 15, no. 1 (2012): 82–87, https://www.cambridge.org/core/journals/public-health-nutrition/article/relationship-between-ultraprocessed-foods-and-metabolic-syndrome-in-adolescents-from-a-brazilian-family-doctor-program/198EA399730DBF2C5488F0F1360C9B6D; W.J. Khalil et al., 'Environmental Pollution and the Risk of Developing Metabolic Disorders: Obesity and Diabetes', *International Journal of Molecular Sciences* 24, no. 10 (2023): 8870, https://www.mdpi.com/1422-0067/24/10/8870; Eurídice Martínez Steele et al., 'Dietary Share of Ultra-Processed Foods and Metabolic Syndrome in the US Adult Population', *Preventive Medicine* 125, (August 2019): 40–48, https://www.sciencedirect.com/science/article/pii/S0091743519301720.

children and adolescents and becoming a major public health challenge worldwide.[15]

The industrial agricultural sector can be defined as one of the principal actors in 'predatory globalisation', which prefers to be ranked on the efficiency of capital, rather than on people's well-being. First of all, this is a political issue, considering that industrial food is produced at high cost, because of public subsidies, yet is marketed internationally through so-called 'free trade treaties'. Local markets, flooded with cheap junk food, lose their suppliers—and farmers, under the pressure of a contrived productive system, are forced to abandon their land.

Food transformation is a process that accounts for about three-fourths of all international food sales. Healthy substances, such as vitamins, are generally removed, and large amounts of sugars and fats, preservatives, organic solvents, hormones, colouring agents, flavour enhancers and other food additives are normally added in the process, especially when the food has to travel thousands of kilometres and must be transformed to increase its shelf-life. The effects of these additives are often unknown, while their interactions with other substances present in food have not yet been identified.

According to the authors of the *Manifesto on Food for Health*,[16] diets rich in calories but poor in fibres and nutrients, together with high fats, sugar and salt, are associated with high blood pressure, high blood sugar, high blood lipids and body fat. These in turn trigger the pathological processes of inflammation, atherosclerosis of blood vessels and thrombosis, and induce carcinogenesis through epigenetic effects.

When we look at living systems as a whole, disease is treated

---

[16] *Manifesto on Food for Health*, op. cit.

Dead Food, Dead Metabolism

as a metabolic disorder. The mechanistic medical paradigm sees individual diseases and looks for cures by suppressing and eliminating the distinct disease. Drugs target a *response* rather than the disease. However, in a systems-health paradigm, different diseases are seen as symptoms of underlying causes: foodstyle/lifestyle-induced metabolic disorders producing mitochondria dysfunction. Metabolic disorders need metabolic strategies. In the context of disease caused by junk food, health can be regenerated by shifting to diverse, chemical-free, fresh or artisanally-processed food.

The ecological health paradigm sees the climate crisis as a disruption of the earth's self-regulating capacity caused by the destruction of the biosphere. Systems that regenerate biodiversity as well as the biosphere also mitigate climate change and create climate resilience. Diversity is the principle for resilience and immunity. Diversity is also the basis of sustainable, healthy and productive food systems.

5 | The Fake Food Dystopia

Through her biosphere and her complex ecological processes, the earth regulates her water systems, her nutrient cycle and her climate system. We have a duty to live within the ecological limits that the earth sets. We do not have the right to pollute and disrupt her ecological processes; taking the share of other species, other people and future generations is an ecological crime.

Half a century of eating oil, two hundred years of industrialisation, combined with the colonisation of nature and of our diverse cultures, have destroyed the earth's ecosystems and biosphere. Fossil fuels have polluted the land, water and atmosphere, built up emissions and disrupted the delicate balance of the earth's climate systems and hydrological systems, intensifying floods, droughts and cyclones, causing climate chaos and catastrophe.

We have seen how corporations like Bayer, Monsanto, Cargill, Coca-Cola, PepsiCo and Nestlé have pushed 93 percent of crop diversity to extinction. Even as industrial agriculture has brought the planet and our economies to the brink of collapse, it is reinventing its future based on fake farming and fake food, with more chemicalisation, more GMOs and more mechanisation, combined with digital agriculture, surveillance drones, robots and spyware. Farming without farmers, without biodiversity, without soil, is the vision of those who are accelerating the ecological collapse.

This industrialised system is currently pushing digital agriculture and fake, farm-free food—laboratory-engineered

The Fake Food Dystopia

meat, milk, cheese, fish and even lab-made breast milk. The health costs, social costs, energy costs and climate costs of this system have not yet been assessed.

The columnist George Monbiot is among the messiahs of fake food. He works with a group that calls itself 'ecomodernists', who have launched a campaign to 'reboot' food as part of 'RePlanet'.[1] Quite clearly, the living earth is inadequate for them and needs to be re-engineered. Mark Lynas, an ecomodernist, also works for the Gates-funded Cornell Alliance for Science, which promotes the genetic engineering of seeds and food. They do not understand that food is not a Microsoft programme that needs to be rebooted.[2] Monbiot writes, 'Lab-grown food will soon destroy farming—and save the planet.'[3] His assumptions are false at every level.

Lab food will not put an end to industrial farming—it will accelerate and expand it. As Bob Reiter, Bayer's head of Research and Development in the company's Crop Science Division, says,

> In order for plant-based companies to produce at scale and succeed, they require efficient sources of amino acid and

---

[1] A campaign which has its roots in a network of ecomodernist groups and societies established since 2015, after the publication of *An Ecomodernist Manifesto*.

[2] Mike Hannis, 'Rebooting Reality', *The Land*, no. 32, https://www.thelandmagazine.org.uk/articles/rebooting-reality; Jonathan Matthews, 'George Monbiot Teams Up With Mark Lynas and the Ecomodernists to Reboot Food', GMWatch, November 14, 2022, https://www.gmwatch.org/en/106-news/latest-news/20127.

[3] George Monbiot, 'Lab-Grown Food Will Soon Destroy Farming—and Save the Planet', *The Guardian*, January 8, 2020, https://www.theguardian.com/commentisfree/2020/jan/08/lab-grown-food-destroy-farming-save-planet.

carbohydrates, which will bring them round to grow crops that can be tilled and cultivated by machinery.[4]

Lab-grown counterparts require massive energy intensive bioreactors and the use of sterile, single-use plastic equipment. To come close to matching current meat consumption, for example, production facilities would need to number in the tens of millions, increasing problematic plastic consumption and escalating energy requirements, all while still relying on globalised industrial agriculture models and supply chains.

Most significantly, in order to function, these bioreactors require large amounts of nutrients for cells to grow and reproduce. Given the limited production of individual amino acid formulations suited for cell culture globally, one hope is to use soya to derive the full amino acid profile necessary for cell growth. This will further entrench the already destructive cultivation of soya.

Gruesomely, and ironically, other parts of the nutrient broth used to culture cells also directly derive from current industrial animal production, as some of them are made using foetal cow's blood obtained from conventionally slaughtered pregnant cows. Stem cells necessary for cell reproduction during the cell culturing process are obtained from foetal cows as well. Without an abundance of slaughtered foetal cows, can cell-cultured meat scale up? And can lab-grown meat solve the problem of animal welfare and environmental degradation if it is completely

---

[4] Tina Bellon, 'Bayer Sees Potential Future Business In Plant-Based Meat Market', Reuters, August 1, 2019, https://www.reuters.com/article/us-bayer-agriculture-food/bayer-sees-potential-future-business-in-plant-based-meat-market-idUSKCN1UR5SF.

dependent on ingredients that derive from industrial beef production? This grim reality says otherwise.

Meat analogs and cell-based meats are much more carbon intensive than we are led to believe. A preprint study, not yet peer-reviewed, by researchers at the University of California, Davis, reveals that the fossil fuel energy required to produce lab meat is not sustainable and could, by far, surpass the output of livestock like pigs and poultry.[5]

The production of synthetic foods, too, requires vast amounts of energy, including several energy intensive steps such as operating bioreactors, temperature controls, aeration and mixing processes. On the basis of these indicators, the sector is in no position to claim that synthetic food production is inherently more sustainable than traditional systems. A recent paper by Purdue University indicates that lab meat will require more acreage and need more production of grain than real beef:

> The total amount of feed excluding harvested forages fed to beef cattle equals approximately 64.5 million tons in 2019 in the U.S. Roughly 57% of this amount came from corn and only roughly 4% was soy-related ingredients.[6]

---

[5] Amy Quinton, 'Lab-Grown Meat's Carbon Footprint Potentially Worse Than Retail Beef', University of California, Davis, May 22, 2023, https://www.ucdavis.edu/food/news/lab-grown-meat-carbon-footprint-worse-beef.

[6] Yanyu Ma, H. Holly Wang, Yizhou Hua, and Shihuan Kuang, 'The Rise of Meat Substitute Consumption and Its Impact on the U.S. Soybean Industry', *Purdue Agricultural Economic Report*, May 15, 2023, https://ag.purdue.edu/commercialag/home/paer-article/the-rise-of-meat-substitute-consumption-and-its-impact-on-the-u-s-soyabean-industry; see also 'The Irony of Protein Corporations', Regenetarianism, July 17, 2023, https://lachefnet.wordpress.com/2023/07/17/the-irony-of-protein-corporations.

The researchers' analysis, based on certain assumptions, goes on to note that plant-based, and especially cell-Ag proteins, increase the market for soyabean, since soya isolates are used in many of the plant-based products and soya hydrolysate is the most available source of amino acids used in the cell media to grow stem cells. They conclude, 'Our major results show that one unit of conventional beef, plant-based beef and cell-cultured beef would require 0.24, 0.54 and 1.44 units of soyabean, respectively, to feed. . .' In other words, plant-based 'meat' and cell-Ag actually increase the demand for soya since these forms of alt protein require, respectively, over two to nearly six times the amount of soya needed for conventional beef.

Lab-grown meat, which is cultured from animal cells, is often thought to be more environmentally friendly than beef because it is said to need less land and water and to emit fewer GHGs than raising cattle. But the UC Davis research shows that lab-grown or 'cultivated' meat's environmental impact is likely to be 'orders of magnitude' higher than retail beef based on current and near-term production methods.[7] The researchers conducted a life-cycle assessment of the energy required and GHGs emitted during all stages of production and compared it with natural beef.

One of the current challenges posed by lab-grown meat is the use of highly refined or purified growth media required to help animal cells multiply.[8] Currently, this method is similar to the biotechnology used to make pharmaceuticals. It sets

---

[7] Derrick Risner et al., 'Environmental Impacts of Cultured Meat: A Cradle-to-Gate Life Cycle Assessment', preprint, submitted April 21, 2023, https://www.biorxiv.org/content/10.1101/2023.04.21.537778v1.full.pdf+html.

[8] Derrick Risner et al., 'Cradle to Production Gate Life Cycle Assessment of Cultured Meat Growth Media: A Comparison of Essential 8™ and Beefy-9',

The Fake Food Dystopia

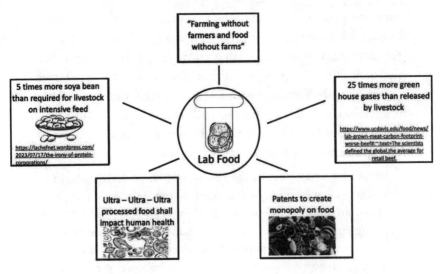

The production of lab food is resource intensive, energy intensive and extractive.

up a critical question for cultured meat production: is it a pharmaceutical product or a food product? 'If companies are having to purify growth media to pharmaceutical levels, it uses more resources, which then increases global warming potential,' says Derrick Risner, lead author and doctoral graduate, UC Davis Department of Food Science and Technology. 'If this product continues to be produced using the "pharma" approach, it's going to be worse for the environment and more expensive than conventional beef production.'

Feedstock intensive and energy intensive lab food, synthetic food and cellular food will increase feedstock demands fivefold and increase emissions by 25 times.[9] The researchers defined the global

---

preprint, submitted April 21, 2023, https://www.biorxiv.org/content/10.1101/2023.04.21.537772v1.

[9] Quinton, op. cit.

warming potential as the carbon dioxide equivalents emitted for each kilogram of meat produced; their study found that the global warming potential of lab-grown meat, using these purified media, is *four to 25 times greater* than the average for retail beef.

Jim Thomas (formerly with ETC), writing in defence of Chris Smaje, a social scientist, farmer and author of the book *Saying NO to a Farm-Free Future* (in which he counters the claims of food solutions based on techno-fixes), says,

> He [Chris Smaje] shows that every kilo of bacterial protein will require at least 65 Kwh of energy—twice the daily energy use of an average US household. This is electricity use which in aggregate would then have to be added on top of expected additional electricity demand for electric vehicles, electric heating of our homes, running an ever-ballooning internet, cloud and AI infrastructure and much more—all from clean energy sources without damaging[,] mining and extraction for the infrastructure build out. Food would seem to be an unnecessary use of additional electricity generation since for now agroecological land-based food production doesn't require electricity at all. Incredibly[,] food really does grow on trees.[10]

The microorganisms used in lab food need to grow and produce commodities; the feedstock for the nutrient path is derived from industrial crops, including genetically modified corn and soyabean. The nutrient conversion efficiency of microbes needs to be compared to livestock, the resource and energy efficiency needs to be compared to animal-based products and the land used for feedstock has to be included while calculating real efficiency, as well as the pollution generated and waste material

---

[10] Jim Thomas, 'George and the Food System Dragon', Scan The Horizon, October 26, 2023, https://www.scanthehorizon.org/p/george-and-the-food-system-dragon.

produced. The infrastructure cost of thousands of fermentation tanks and the particular environmental conditions required for microorganisms to grow and thrive need to be internalised in any cost accounting of lab foods. All these issues were ignored in factory farming and are now being ignored for lab substitutes.[11]

Thomas also points out that in order to grow, lab-produced meat, just like animals on farms, would need

> ongoing nutrient addition of phosphate and nitrogen as well as other elements—which must be acquired, mined, carried[,] etc with additional energy and biodiversity costs. Of course[,] industrial agriculture already requires large amounts of these artificially acquired inputs but not agroecological farming[,] which draws nutrients from soil and animal waste.[12]

In other words, lab meat will continue to be resource intensive and extractive.

Another potentially serious problem in lab meat production or alt protein, as Thomas points out, is the high risk of contamination, since fermentation processes attract wild bacteria and yeasts. Lab food, combined with further industrialisation of farming, intensifies all the processes that have contributed to the agrarian crisis, the climate crisis and the health crisis.

While further industrialisation and globalisation of food in the form of fake food (as a false solution to the climate crisis) are being assiduously promoted, carbon dioxide emissions resulting from the industrial processing of synthetic meat may

---

[11] Errol Schweizer, 'What Consumers Should Ask about Precision Fermentation', *Forbes*, March 2, 2022, https://www.forbes.com/sites/errolschweizer/2022/03/02/what-should-consumers-be-asking-about-precision-fermentation.

[12] Thomas, op. cit.

persist in the atmosphere for hundreds of years, unlike methane produced by traditional livestock farming, which dissolves in the atmosphere after about ten years.[13] Therefore, synthetic food does not seem to have a legitimate place in the category of eco-friendly food; rather, it belongs in the category of ultra-processed food, due to the high-impact transformation process required for its production.

The idea of farming without farmers and food without farms is dystopian, a continuation of the false assumption of eco apartheid—that we are separate from nature and can live outside the earth's life-giving processes.

Continuing on the resource intensive, energy intensive, pollution and emissions intensive path of the industrial food system while expecting a different outcome is a clear sign of insanity; as Einstein had said, 'Insanity is doing the same thing over and over and expecting different results.'[14]

Fake food and who benefits

The Poison Cartel, Big Food and Big Money are investing millions of dollars in the fake food industry to support the mass proliferation of fake food products such as eggs, dairy and meat. Most fake food ventures are funded by tech oligarchs

---

[13] Samantha Werth, 'The Biogenic Carbon Cycle and Cattle', CLEAR Center at UC Davis, February 19, 2020, https://clear.ucdavis.edu/explainers/biogenic-carbon-cycle-and-cattle.

[14] Vandana Shiva, 'Rewilding Food, Rewilding Farming', *The Ecologist*, January 24, 2020, https://theecologist.org/2020/jan/24/rewilding-food-rewilding-farming.

[15] For more, see Vandana Shiva, with Kartikey Shiva, *Oneness vs. the 1%: Shattering Illusions, Seeding Freedom* (White River Junction, VT: Chelsea Green, 2020).

and billionaires. Bill Gates became a billionaire by enclosing the commons of software,[15] and is planning the same enclosure for food. The key questions are: who holds the patents on lab food? What are the implications of one company owning the formula for milk, honey, eggs, meat and our daily bread?

It is no coincidence that, along with industrial giants like Cargill and Tyson Foods, even the biggest 'environmental' philanthropists are investing in this sector. High-profile Big Tech investors such as Bill Gates, Amazon founder, Jeff Bezos and Virgin founder, Richard Branson, have joined up by providing substantial financial support to start-ups and biotechnology companies pursuing innovations in the fake food sector. Gates alone has already invested US$50 million in Impossible Foods and actively finances Beyond Meat, Ginkgo Bioworks, BIOMILQ, Motif FoodWorks, C16 Biosciences, Hampton Creek Foods and Memphis Meats (now UPSIDE Foods) through his Breakthrough Energy Ventures investment fund.[16]

Other prominent start-ups funded by this billionaire investment fund include Eat Just (egg substitutes made from plant proteins), Perfect Day (lab-grown dairy products) and NotCo (plant-based animal products made through AI). Indeed, the promotion of fake foods seems to have more to do with giving new life to the failing GMO agriculture and junk food industry than to provide a solution to the food crisis,

---

[16] Anna Starostinetskaya, 'Jeff Bezos, Bill Gates, and Richard Branson Lead $90 Million Investment to Create Next Vegan Impossible Burger', *VegNews*, February 26, 2019, https://vegnews.com/2019/2/jeff-bezos-bill-gates-and-richard-branson-lead-90-million-investment-to-create-next-vegan-impossible-burger; see also Zack Friedman, 'Why Bill Gates and Richard Branson Invested in "Clean" Meat', *Forbes*, August 25, 2017, https://www.forbes.com/sites/zackfriedman/2017/08/25/why-bill-gates-richard-branson-clean-meat.

as well as countering the threat from a rising consciousness regarding organic, local, fresh food that regenerates the planet. Consequently, investment in plant-based food companies soared from nearly nil in 2009 to US$600 million in 2018.

Over the last couple of years, and following the rapid growth of new start-ups, the market for synthetic and plant-based alternatives has been expanding, with financial backing skyrocketing in 2020. The Good Food Institute, a lobby group for the adoption of animal product alternatives, reports that in the United States, the plant-based market has already grown from US$4.9 billion in 2018 to US$7 billion in 2020, which represents an overall increase of 43 percent in dollar sales over the two years. Plant-based meat, too, is booming, having reached a value of US$1.4 billion and registered a growth of 72 percent in 2020.[17] The synthetic biology industry follows close behind; it reached a value of US$12 billion in the last decade, is expected to double by 2025, reaching US$85 billion by 2030. Companies specialising in this field have grown sixfold over the last ten years. This exponential growth is confirmed by the recent figures from SynBioBeta (a network of biological engineers, investors, innovators and entrepreneurs): the first quarter of 2021 saw record investments of US$4.7 billion in start-ups, and of US$4.2 billion in the second quarter.[18] Over the last two decades, the

---

[17] Kyle Gaan, 'Plant-Based Food Retail Sales Reach $7 billion', Good Food Institute, April 6, 2021, https://gfi.org/blog/spins-data-release-2021/; Brian Kateman, 'Healthier Plant-Based Meat Is on the Rise', *Forbes*, May 10, 2021, https://www.forbes.com/sites/briankateman/2021/05/10/healthier-plant-based-meat-is-on-the-rise.

[18] 'Q1 Shatters Previous Synthetic Biology Investment Record–Signals Projected 2021 Investment of up to $36 Billion', SynBioBeta, April 7, 2021, https://www.synbiobeta.com/read/q1-shatters-previous-synthetic-biology-investment-record-signals-projected-2021-investment-of-up-to-36-billion.

number of companies specialising in this field has risen from less than 100 in 2000 to more than 600 in 2019. Beyond Meat was one of the 'hottest' stocks in 2019, its shares growing a phenomenal 859 percent in the first three months of its incorporation!

Among the top players in the fake food sector are Mosa Meat, Eat Just, Inc., JUST Egg, GOOD Meat, MeaTech 3D, Aleph Farms, CUBIQ FOODS, Because Animals, BlueNalu and UPSIDE Foods.[19] The last-mentioned has attracted the attention of Richard Branson, Suzy and Jack Welch, Kimbal Musk and Bill Gates. This company uses biotechnology to extract and cultivate stem cells in different muscle tissue, which is then fed into bioreactors in order to grow meat products.

Fake chicken uses a bioreactor which accelerates the growth of animal cells taken from poultry. Eat Just produces lab-grown chicken, for which it received sales approval in Singapore in 2020. Founded in 2011 in San Francisco by Josh Tetrick, Eat Just also produces plant-based egg product alternatives. In 2016, it surpassed US$1 billion in valuation, becoming a unicorn in the protein foods market. MeaTech and Aleph Farms both use the 3D printing technology to develop what they claim are sustainable alternatives to conventional meat. As for CUBIQ FOODS, it specialises in providing the fatty ingredients needed to improve the taste of fake meat. Because Animals provides a niche product by way of lab-grown meat for pets. The seafood sector is occupied by BlueNalu which is into the large-scale commercialisation of cell-cultured seafood. It is developing lab-grown counterparts of bluefin tuna, mahi-mahi, red snapper, molluscs, crustaceans and other seafood.

---

[19] Joseph Mapue, '8 Pioneering Companies Creating Sustainable Lab-Grown Meat', Ross Dawson, https://rossdawson.com/futurist/companies-creating-future/8-pioneering-companies-creating-sustainable-lab-grown-meat.

The corporate push for synthetic foods.

(*Source: Navdanya International*)

A German start-up company, Formo, has received record funding of US$50 million from its shareholders for the production of ricotta and mozzarella in laboratories. The funding represents a record for a European foodtech start-up and sends a clear signal to investors and markets around the world. 'Founded in 2019, Formo is Europe's first cellular agriculture company to develop animal-free dairy products using real, nature-identical milk proteins derived from precision fermentation.' [20]

---

[20] Thomas Ohr, 'Berlin-Based Foodtech Startup Formo Raises €42 Million to Supercharge Animal-Free Cheese Production', EU-Startups, September 13, 2021, https://www.eu-startups.com/2021/09/berlin-based-foodtech-startup-formo-raises-e42-million-to-supercharge-animal-free-cheese-production.

The Fake Food Dystopia

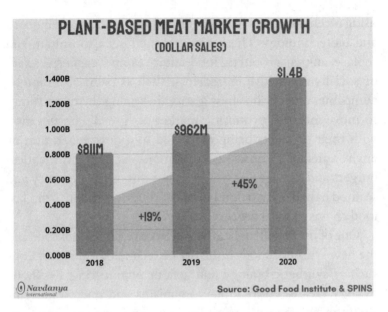

Over the last few years, the market for plant-based foods has been steadily expanding.

(Source: Navdanya International)

Given the widespread success of the plant-based industry, it is not surprising that big plant-breeding companies like Bayer see a great opportunity for investment and expansion in this market. Clearly, it is the Poison Cartel, the ultra-processed junk food industry and agribusiness that stand to profit from this lucrative and rapidly expanding market.

Wooing green consumers

The creation of so many start-ups in the fully artificial food sector indicates the increasing popularity of developing a line of synthetically-produced, ultra-processed food products by using recent advances in synthetic biology, artificial intelligence and biotechnology. These new products seek to imitate and replace animal products, food additives and expensive, rare or socially conflictive ingredients (such as palm oil). Biotech companies and agribusiness giants are seizing the opportunity to move into this promising market of 'green' consumption with their products being marketed to a new generation of environmentally conscious consumers. As a result, meatless burgers and sausages, as well as imitations of cheese, dairy and seafood have begun to flood markets, found anywhere from fast food chains to local grocery stores.

One of the key differences between conventional junk food and the new synthetic foods is the use of technological innovations such as synthetic biology and genetic engineering. Synthetic biology creates entirely new organisms and microorganisms through the genetic modification or engineering of an organism's internal genetic components to reconfigure them in new ways. By implanting bits of other organisms' DNA into microorganisms, or reconfiguring internal genetic information, these new technologies induce microorganisms, cells or other genetic material to 'ferment' and reproduce in order to create new, completely synthetic ingredients. The use of the word 'fermentation' in synthetic biology creates a false analogy between traditional forms of natural microbial fermentation and these new, completely artificial biotechnologies.

These technologies are being used by companies such as Beyond Meat, Motif FoodWorks, Ginkgo Bioworks (custom-built microbes), BIOMILQ (lab-grown breast milk), Nature's Fynd (fungi-grown meat and dairy alternatives), Eat Just (egg substitutes made from plant proteins), Perfect Day (lab-grown dairy products) and NotCo.

The Fake Food Dystopia

Companies like Beyond Meat and Impossible Foods use a DNA coding sequence derived from soyabean or peas to create a product that looks and tastes like real meat. Filler ingredients for these products still rely heavily on the extensive processing of conventionally-cultivated and mostly GMO crops. For instance, the Impossible Burger is made almost entirely from industrially produced wheat, maize, soya, coconut and potato, as well as additional bioengineered ingredients. Proteins and carbohydrates from these conventional crops are chemically extracted, cooked, and then extruded through machines that blend and shape them into strands resembling short muscle fibres, allowing manufacturers to convincingly imitate a range of processed meat products.

In cell-based meat, tissue is taken from a live cow and combined with extracted stem cells to grow into muscle fibre in the lab. Once enough (over 20,000) have been obtained from this process, they are coloured, minced, mixed with fats and shaped into burgers.

UPSIDE Foods produces meat by using self-reproducing animal cells. The rationale is that such a process eliminates the need to breed and slaughter larger numbers of animals, thus side-stepping ethical and ecological concerns along the supply chain. While lab-grown meat is not yet widely available to the public, companies like UPSIDE Foods are investing heavily in research and development in order to make their products economically affordable over the long term and to compete with commercial meat.

Whether upscaling lab-grown meat will one day be economically viable remains doubtful, especially since there are many obstacles faced by cultured meat companies. Scientific data demonstrates that cultivated meat gives rise to many inefficiencies and limitations in scalability, as evident in the need for intensive and sophisticated machinery, structural limits on cell metabolism and immunity to foreign contaminants and a series of complex processes that place a strict limit on the expansion of production. All these contribute to a lack of cost competitiveness with the conventional meat products they wish to replace, especially as cell-culturing facilities at the scale needed have never been made viable.[21]

Being energy, resource and capital intensive, the lab food and fake food economy is highly non-sustainable. In fact, dependency on capital is so high that despite doing very well for six years, the UK company Meatless Farm was forced to lay off much of its workforce as its biggest investors and potential new investors decided to discontinue funding due to a reported slump in plant-based meat sales along with increasing competition.[22] Post-Covid, some outlets, including Hopdoddy Burger Bar, preferred to turn to regenerative meat and in-house vegetarian patties.[23]

---

[21] Jean-François Hocquette et al., 'Review: Will "Cultured Meat" Transform Our Food System Towards More Sustainability?', *animal* (2024), https://www.sciencedirect.com/science/article/pii/S1751731124000764; Xin Li Ching et al., 'Lab-Based Meat the Future Food', *Environmental Advances* 10 (December 2022), https://www.sciencedirect.com/science/article/pii/S2666765722001508; Konstantina Kyriakopoulou et al., 'Plant-Based Meat Analogues', in Charis M. Galanakis, ed., *Sustainable Meat Production and Processing* (Cambridge, MA: Academic Press, 2019), 103–26, https://doi.org/10.1016/B978-0-12-814874-7.00006-7.

[22] Maxwell Rabb, '"Nobody Expected This": Beloved Vegan Meat Brand Prepares for Bankruptcy', Plant Based News, June 20, 2023, https://plantbasednews.org/news/economics/vegan-meat-brand-meatless-prepares-for-bankruptcy.

[23] Bret Thorn, 'Hopdoddy Burger Bar Adds More Regenerative Meat to Its Menu As it Removes Manufactured Plant-Based Protein', *Nation's Restaurant News*, September 19, 2023, https://www.nrn.com/fast-casual/hopdoddy-burger-bar-adds-more-regenerative-meat-its-menu-it-removes-manufactured-plant.

The Fake Food Dystopia

In an article on the Impossible Burger, Pat Brown, CEO and founder of Impossible Foods, says, 'We sought the safest and most environmentally responsible option that would allow us to scale our production and provide the Impossible Burger to consumers at a reasonable cost.'[24]

The Impossible Burger, based on vast monocultures of GMO and Roundup-sprayed soya, cannot be considered a 'safe' option,

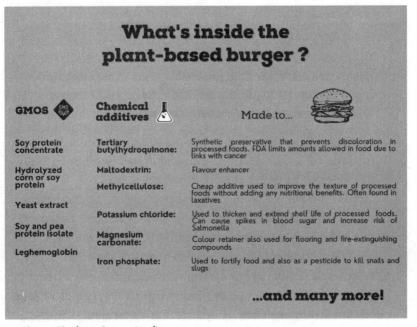

(Source: Navdanya International)

[24] Pat Brown, 'How Our Commitment to Consumers And Our Planet Led Us to Use GM Soy', Medium, May 16, 2019, https://medium.com/impossible-foods/how-our-commitment-to-consumers-and-our-planet-led-us-to-use-gm-soy-23f880c93408.

both because of its high levels of glyphosate and its effect on our gut microbiome. Zen Honeycutt of Moms Across America says,

> The levels of glyphosate detected in the Impossible Burger by Health Research Institute Laboratories were 11x higher than the Beyond Meat burger...This new product is being marketed as a solution for 'healthy' eating, when in fact 11 ppb of glyphosate herbicide consumption can be highly dangerous.[25]

Pat Brown admits:

> We use genetically engineered yeast [obtained from soya DNA] to produce heme, the 'magic' molecule that makes meat taste like meat—and makes the Impossible Burger the only plant-based product to deliver the delicious explosion of flavor and aroma that meat-eating consumers crave.[26]

In fact, the Impossible Burger is a plant-based burger, the key ingredient of which is a protein called soya leghemoglobin (SLH), derived from genetically modified yeast.

Artificial meat is also made up of protein and fat from peas, potatoes, soya and maize grown in monocultures, based on the same heavy-duty processing methods, chemical inputs and GMOs that compromise global biodiversity, destroy wildlife, alter soils and pollute groundwater sources. Yet, the first benefit highlighted in the marketing campaigns of synthetic food companies remains that of reduced environmental impact. This assumption is difficult to prove, considering that plant-

---

[25] Zen Honeycutt, 'GMO Impossible Burger Positive for Carcinogenic Glyphosate', Moms Across America, July 8, 2019, https://www.momsacrossamerica.com/gmo_impossible_burger_positive_for_carcinogenic_glyphosate.

[26] Pat Brown, op. cit.

based synthetic foods are based on exactly the same system as industrial agriculture. And 'plant-based' is often shorthand for 'plant-based proteins', meaning proteins derived from plants rather than from meat or dairy.

The health hazards of industrially- and ultra-ultra-processed foods are widely recognised, and the latest generation of junk synthetic foods are no better; in fact, they are worse, since genetically engineered artificial ingredients as well as chemically extracted protein isolates are used to produce them. They are as detrimental to health as everyday ultra-processed foods because they contain the same health-altering ingredients: additives, fat, sodium, sugar. With regard to the controversial SLH produced through synthetic biology, according to the Center for Food Safety, the Food and Drug Administration (FDA) did not conduct adequate long-term testing before approving the additive in 2019; after a short-term rat trial, several potential adverse effects were detected such as weight gain, changes in the blood that

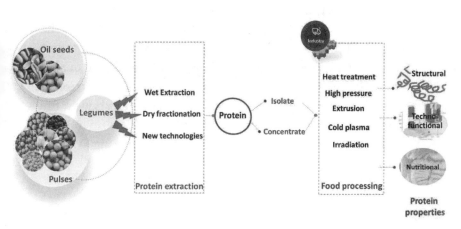

Steps involved in plant-based food processing.

(Source: https://www.mdpi.com/2227-9717/10/12/2586)

could indicate inflammation or kidney disease, disruptions in the reproductive cycle and possible signs of anaemia.[27]

Despite this lack of evidence regarding its safety, Impossible Foods' products containing genetically engineered heme are being sold in supermarkets across the United States, exemplifying a reprehensible laxity in testing and regulation for these new products and technologies.

The Canadian company BetterMilk, is investing heavily in the production of cow's milk using bovine mammary cells. TurtleTree, a start-up lab based in Singapore and the US, is poised to launch human lactoferrin in the market as the first commercial cellular product for newborns. US-based BIOMILQ has also announced that it is ready to market the first synthetic baby milk cultured from human cells. Is this product comparable to breast milk? The company doesn't claim so but says, 'It will be free from the environmental toxins, food allergens, and prescription medications that are often detected in breast milk,' because 'it is produced outside the body in a controlled, sterile environment.' Indeed, this is a well-founded argument, as the latest studies on natural breast milk show. A recent American intercollegiate study found the presence of PFAS (a group of synthetic chemicals used in consumer products) in 100 percent of breast milk analysed: all 50 samples examined showed the presence of dangerous chemical substances at levels up to 2,000 times higher than those considered safe in drinking water.[28]

---

[27] 'Rat Feeding Study Suggests the Impossible Burger May Not Be Safe to Eat', Organic Consumers Association, April 1, 2023, https://organicconsumers.org/rat-feeding-study-suggests-the-impossible-burger-may-not-be-safe-to-eat; originally published in GMWatch, September 20, 2022, https://www.gmwatch.org/en/106-news/latest-news/20099-rat-feeding-study-suggests-the-impossible-burger-may-not-be-safe-to-eat.

BREAST MILK CULTURED FROM HUMAN CELLS: BIG BUSINESS' NEW CONQUEST

PFAS IN BREAST MILK

100% OF AMERICAN MOTHERS ARE VICTIMS OF ENVIRONMENTAL POLLUTION

A recent American intercollegiate study found the presence of PFAS in 100% of the breast milk analysed: all 50 samples examined showed the presence of the dangerous chemical substances at levels up to 2,000 times higher than those considered safe in drinking water.

THE SOLUTION FROM THE INDUSTRY

AVOID POLLUTION? NO NEED. THERE'S BIO-MILK!

According to US-based Biomilq, which is poised to market the first synthetic baby milk grown from human cells, bio-milk is superior to human milk because it is 'produced outside the body in a sterile, controlled environment, free of the environmental toxins, food allergens and prescription drugs that are often present in breast milk'.

THE USUAL SUSPECTS ARE ON THE MOVE

BILL GATES BETS ON THE SECTOR

Biomilq has announced the closing of its Series A financing round with $21 million. The North Carolina-based biotech attracted funding from existing investor Breakthrough Energy Ventures, a company founded by Bill Gates. Other participating investors included Blue Horizon, Spero Ventures, Digitalis Ventures, Alexandria and Gaingels. Green Generation Fund, Europe's first female-led fund, also invested in Biomilq.

(Source: Navdanya International)

Yet, in spite of this, the industrial apparatus systematically aims to invest in technological solutions in order to profit from the problems it has itself created.

Synthetic and lab foods represent yet another profit-making machine used by billionaires and big corporations to capitalise on proprietary technology, reflected in companies' relentless pursuit of patents for anything from novel synthetic biology processes to genetically engineered ingredients like SLH and protein texturising processing and even the patenting of genetic material used as raw material. As we demonstrated in our report *Gates to a Global Empire*,[29] 27 patents have been

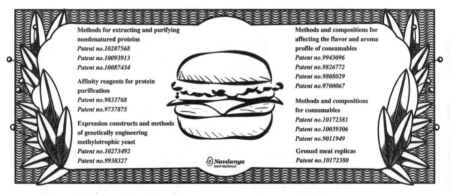

A SOFTWARE TO SWALLOW: PATENTS FILED FOR THE ARTIFICIAL HAMBURGER

(Source: Navdanya International)

---

[28] '100% of Breast Milk Samples Tested Positive for Toxic "Forever Chemicals"', Toxic-Free Future, May 13, 2021, https://toxicfreefuture.org/press-room/100-of-breast-milk-samples-tested-positive-for-toxic-forever-chemicals/; see also Guomao Zheng et al., 'Per- and Polyfluoroalkyl Substances (PFAS) in Breast Milk: Concerning Trends for Current-Use PFAS', *Environmental Science & Technology* 55, no. 11 (2021): 7510–20, https://doi.org/10.1021/acs.est.0c06978.

The Fake Food Dystopia

assigned to Impossible Foods, with over one hundred additional patents pending for other fake meat proxies, from chicken to fish. The patenting logic that underlies the synthetic food movement sees animals and nature as disposable elements that can simply be replaced by more efficient technologies and lab-engineered products.

The EAT forum and Big Food

The Poison Cartel and the junk food industry promote fake food in partnership with the EAT forum, which is closely associated with the World Economic Forum (WEF). EAT has a partnership through FReSH (Food Reform for Sustainability and Health) with the junk food industry and Big Ag, such as Bayer, BASF,

EAT forum's FReSH initiative has partnered with over thirty companies to 'transform' the food system.

(Source: 'FReSH,' EAT, https://eatforum.org/initiatives/fresh/)

---

[29] Carla Ramos Cortés, *Gates to a Global Empire...over Seed, Food, Health, Knowledge and the Earth—A Global Citizens' Report*, Navdanya International, October 2020, https://navdanyainternational.org/publications/report-synthesis-gates-to-a-global-empire.

Cargill and PepsiCo, among others. Bayer became the biggest GMO seed and agrochemical company after it merged with Monsanto, and Yara is the biggest chemical fertiliser corporation in the world.

In 2019, the EAT forum released the EAT–*Lancet* Report, 'Food in the Anthropocene: The EAT–*Lancet* Commission on Healthy Diets from Sustainable Food Systems', that sought to impose a monoculture diet of chemically-grown, hyper-industrially-processed food on the world, claiming that a 'healthy and sustainable [plant-based] diet' protects the health of the planet and of people. By not addressing the key danger of pesticides and toxins in growing and processing food in chemical intensive industrial agriculture, the report turned a blind eye to chronic disease epidemics in many parts of the world.

What is cause for real alarm is that synthetic food is slowly encroaching on multilevel governance arenas. This was most evident at the 2021 UN Food Systems Summit (UNFSS), as well as at COP 26 in Glasgow, Scotland, both of which served as forums to reveal the true intention of agribusiness and food giants—keeping the system unchanged. Both summits were yet another failed attempt at addressing power imbalances in the food system, with sustainable farming practices like agroecology playing a marginal role.

Reflected in the themes and proposals in both international events was a willingness to maintain business as usual—relying on the failed industrial agricultural model and allowing big actors to dictate terms. During both the UNFSS (September 2021) and COP 26 (October–November 2021), there was an explicit promotion of artificial and ultra-processed plant-based foods, cloaked in the language of achieving 'protein diversification' and 'sustainable diets'. During the UNFSS in New York, Action Track

The Fake Food Dystopia

2, 'Shift to Sustainable and Healthy Consumption Patterns'—led by Nestlé, Danone and the controversial EAT forum—was based on solutions whose sustainability is questionable, and at COP 26, the Plant Based Treaty was promoted and backed by all the abovementioned actors.

In fact, the Bill Gates–run UNFSS, held amid protests from international environmental associations, was where the intentions of the food multinationals found their best expression. The development model must remain the same, namely that of the failed but profitable Green Revolution. The actors involved, i.e., big investors and agribusiness multinationals, must also remain the same in order to continue to profit from new technological investments. What needs to be radically changed is, quite simply, the *narrative*. This is the only truly green element to be found in the action plans of the masters of food! Yet, it cannot be said that we had not been warned. After all, Action Track 1 and 2 of the summit laid out the global strategy very thoroughly.

Action Track 1, 'Ensure access to safe and nutritious food for all', called for large-scale food fortification as a solution to malnutrition. Food-fortification is the process of supplementing food with additional nutrients, which can also involve the use of biotechnology and genetic modification. It is an approach often recommended and put into practice in developing countries where nutritional deficiencies are common.

A classic example is that of Golden Rice, rice that has been genetically modified to contain levels of beta-carotene that can remedy vitamin A deficiencies in the population. The Bill & Melinda Gates Foundation has so far given US$28 million to fund Golden Rice, in direct collaboration with the Global Alliance for Improved Nutrition (GAIN, founded in 2001 by Gates). GAIN, leader of Action Track 1, was among the first organisations to operationalise the public-private partnership model. Since then, it has continued to support biofortification projects to combat malnutrition and food insecurity. GAIN shares many of the same donors as AGRA (Alliance for a Green Revolution in Africa), such as the Rockefeller Foundation, BSF or Unilever, and received no less than US$251 million from the Gates Foundation between 2002 and 2014.

We present here an Impossible Menu of Fake Food: a plate of fortified Golden Rice as the first course, of Impossible Burger as the second course, with a side dish of synthetic mozzarella and vegetables grown from genetically modified seeds. Of course, selling a menu like this isn't very easy, but if stated clearly that such a menu is necessary to protect the environment and that it is also beneficial to our health, perhaps more people would be willing to go along. There may even be more governments willing to fund private research in this artificial nutrition. There is no doubt that a menu conceived in a laboratory, yet at the same time presenting itself as ecologically sound, makes for a seductive narrative. But is this a realistic representation, or are we faced with yet another greenwashing operation to hide the usual suspects behind a fluorescent green veneer? How many of those opting for the Impossible Burger or Beyond Meat are actually informed about the long chain of production, starting from GMO soya to animal parts, of their *green* burger? One thing can be said with certainty: the development of the artificial food industry as the best response to environmental challenges is based on the global food industry trying to reshape its range of products to appeal to an increasingly *green* consumer base, and it is aware of the fact that many of these consumers are not au fait with the causes of the current environmental disasters. It is not surprising that the tremendous rise of synthetic foods is taking

IMPOSSIBLE MENU

GOLDEN RICE

Genetically modified rice containing high levels of beta carotene to remedy Vitamin A deficiencies in the population. The approach based on fortifying one single food rather than increasing the variety and quality of foods available undermines the ability of communities to strengthen local food systems thereby destroying food sovereignty and nutritional biodiversity.

IMPOSSIBLE BURGER

A Hamburger made almost entirely from common crops: wheat, maize, soya, coconut and potato. It contains GMOs. Recent research has found glyphosate levels of 11.3 ppb. This is more than enough to have a negative impact on our intestinal microbiota and therefore on our immune system, not to mention its 'probable' carcinogenicity and its ability to act as an endocrine disruptor.

FAKE MOZZARELLA

A German start-up by the name of Formo has just received a record $50 million in funding from its shareholders to develop large-scale production of lab grown ricotta and mozzarella. The funding represents a record for a European foodtech start-up.

GMO STRAWBERRIES

US companies Simplot and Plant Science Inc. have announced a strategic partnership to bring the first genetically modified strawberry through the CRISPR-Cas9 technology to supermarkets.

(Source: Navdanya International)

THE DANGERS OF FAKE FOOD

Relies on technical innovations such as synthetic biology, which Involves reconfiguring the DNA of an organism to create something entirely new, not found in nature.

Even though fake food is promoted as climate-friendly, its production processes are energy-intensive, and still heavily rely on fossil fuels energy systems, which do not represent an improvement in terms of mitigating the climate impact.

Even though fake food is advertised as "eco-friendly", it is made with proteins from pea, soy, or corn that are grown on large monocultures which still rely on heavy tillage, chemical inputs, and GMOs.

Is a profit-making machine used by billionaires and big corporations to increase control through patents. This logic sees food as a mere commodity up for grabs, endangering the food sovereignty of local communities and small farmers.

Fake food ignores our relationship with nature. It concentrates the power into the hands of a few private companies, and away from indigenous people and farmers, therefore hijacking local food systems and further invalidating their ancestral knowledge.

(Source: Navdanya International)

The Fake Food Dystopia

place at a time when ethical concerns linked to the meat and dairy industry are increasingly in the spotlight. As the industrial agrifood industry is threatened by consumer awareness, big companies that stand to lose significant profits are trying to tap into a new market of environmentally-aware consumers looking for alternatives. It would seem that Big Food has found a way to impose yet another set of technological 'solutions' to a series of problems caused by the very model of industrial agribusiness on which it is based.

If today, ordering a whole meal of fake food at a restaurant seems far-fetched, it might, in the near future, even sound like a responsible choice. In the more distant future, it may no longer be a choice, but the only option left for not leaving the table hungry. If agroecology and organic production are not adequately supported, that future, which seems dystopian right now, may actually come to pass due to a lack of alternatives.

Some might object: but aren't the European Union's Farm to Fork and Biodiversity strategies asking us, among other things, to expand the area under organic production by 25 percent and cut pesticide use by 50 percent by 2030? The answer lies in the implementation of these strategies, which should primarily take place through a reform of the EU's Common Agricultural Policy (CAP) which, when allocating resources, prefers to subsidise large-scale conventional producers. Finally, we should not forget that the Farm to Fork strategy itself inclines towards new technologies of genetic manipulation, which includes the generation of new GMOs. All this notwithstanding, the fact remains that the EU's policies, by far, are the most progressive globally.

If this is the scenario in global politics, what is happening at the individual level? Consumer choices have a major impact on markets, as demonstrated by the growth of organic foods worldwide. According to the International Federation of Organic Agriculture Movements (IFOAM), 2020 recorded the highest growth in the global market for organic products, increasing to €120 billion. Consumers instinctively mistrust synthetic food and this is perhaps why it is necessary to associate these products with an ecological narrative. While vegetarian and vegan diets have the potential for a positive impact on the environment, artificial meat, eggs and cheese substitutes may not. On the contrary, as shown earlier in the chapter, there is much to suggest that the biotech industry is not as sustainable as it claims to be. A growing number of consumers genuinely want to make greener food choices, and if not misled by the industry's greenwashed narrative, would most likely opt for an organic diet. Large investments in the biotech and synthetic food industries could actually delay those regenerative and truly sustainable processes that are trying to emerge, regardless of great difficulties, at a local level all over the world.[30]

Animal-free farming?

Ever since the rise of industrial agriculture, removing farmers from the land has been one of its main objectives. Chemicals and machines have been designed to displace farmers, and now climate change is becoming the driving force for removing them and their animals from the land. 'Animals,' says Pat Brown, 'have just been the technology we have used up until now to produce meat. . . . What consumers value about meat has nothing to do

---

[30] Manilo Masucci, 'An Impossible Menu: Fake Food Is Taking Over Our Tables', *Terra Nuova*, February 2022, https://www.terranuovalibri.it/fascicolo/dettaglio/terra-nuova-febbraio-2022-9788866817000-236614.html.

The Fake Food Dystopia

with how it's made. They just live with the fact that it's made from animals.'

'If there's one thing we know,' argues the Impossible Foods CEO, 'it's that when an ancient, unimprovable technology encounters a better technology that is continuously improvable, it's just a matter of time before the game is over.' Brown also seems on his game when he says that his investors clearly see 'a $3 trillion opportunity' on the horizon.

Ecological sciences have been based on a recognition of the interconnections and interrelatedness between humans and nature, between diverse organisms and within all living systems, including the human body. Diets have evolved according to climates and the biodiversity that the local climate allows. The biodiversity of the soil, of plants and of our gut microbiome are one continuum. Technologies are tools. Tools need to be assessed on ethical, social and ecological criteria. Tools and technologies have never been viewed as self-referential in living cultures; they have been assessed in the context of contributing to the well-being of all.

Through fake food, evolution, biodiversity and the web of life are being redefined as an 'ancient, unimprovable technology', ignorant of the sophisticated knowledges that have evolved in diverse agricultural and food cultures, in diverse climate and ecosystems, to sustain and renew biodiversity, the ecosystem, the health of people and of the planet. Fake food is thus building on a century-and-a-half of food imperialism and food colonisation and on half-a-century of a fossil fuel-based industrial food system.

Animals are being exterminated based on a pseudo-science that says animals are a source of methane pollution. Methane is part of the biogenic cycle of nature. Samantha Werth, senior director of sustainability at the National Cattlemen's Beef Association, USA, explains the process:

> As a by-product of consuming cellulose, cattle belch methane, there-by returning that carbon sequestered by plants back into the atmosphere. After about ten years, that methane is broken down and converted back to CO_2. Once converted to CO_2, plants can again perform photosynthesis and fix that carbon back into cellulose. From here, cattle can eat the plants and the cycle begins once again. In essence, the methane belched from cattle is not adding new carbon to the atmosphere. Rather, it is part of the natural cycling of carbon through the biogenic carbon cycle.[31]

Thirteen countries have joined something called the Global Methane Hub, dedicated to the reduction of methane in various ways. They are: the United States, Argentina, Australia, Brazil, Burkina Faso, Chile, Czech Republic, Ecuador, Germany, Panama, Peru, Uruguay and Spain. US Special Presidential Envoy for Climate, John Kerry, said in a statement,

> Mitigating methane is the fastest way to reduce warming in the short term. Food and agriculture can contribute to a low-methane future by improving farmer productivity and resilience. We welcome agriculture ministers participating in the implementation of the Global Methane Pledge.

Kerry makes no mention of fossil chemicals and nitrous oxide. Targeting methane, which is part of the biogenic cycle, is like targeting animals that have the potential to be partners in ecological agriculture. The problem is not the cow or other

---

[31] Werth, op. cit.

The Fake Food Dystopia

farm animals; the problem is factory farming.

All ecosystems have plants and animals, and all non-violent ecological agriculture systems are based on a symbiosis between them—plants feed animals; animals feed plants. Cows eat plants; plants are nourished by the manure the cows provide, creating a regenerative circular economy that needs no fossil fuel or external inputs and creates no pollution, no waste. Indigenous varieties of crops were bred to maximise both straw and grain, with the straw feeding animals, grains feeding humans. Animals, too, were bred for diversity and were multipurpose, giving food for the soil as manure, energy for farming and food for human consumption.

Industrial agriculture broke this symbiotic relationship between plants and animals. Instead of plants being fertilised with organic manure, they were fertilised with synthetic fertilisers. Plants from indigenous seeds could not tolerate synthetic fertiliser, so they were engineered to be dwarf varieties by Norman Borlaug. Straw, which is food for animals, disappeared. It was substituted with feed made from grain, especially GMO corn and soyabean—40 percent of the corn grown is used for animal feed, 77 percent soyabean is used as feed for livestock, aggravating the hunger crisis. Cows are herbivores and like to eat grass. Shifting them to an intensive grain diet disturbs their metabolism, contributing to increased methane emissions.[32]

The biogenic cycle of renewal is in fact part of a circular

---

[32] Jan Dijkstra, 'Changing The Cow's Diet Reduces Methane and Nitrogen Emissions', Wageningen University & Research, February 16, 2021, https://www.wur.nl/en/article/changing-the-cows-diet-reduces-methane-and-nitrogen-emissions.htm; see also Jena Wilson, 'Reducing the Carbon Footprint of Cattle Operations through Diet', Institute of Agriculture and Natural Resources, University of Nebraska–Lincoln, August 8, 2019, https://

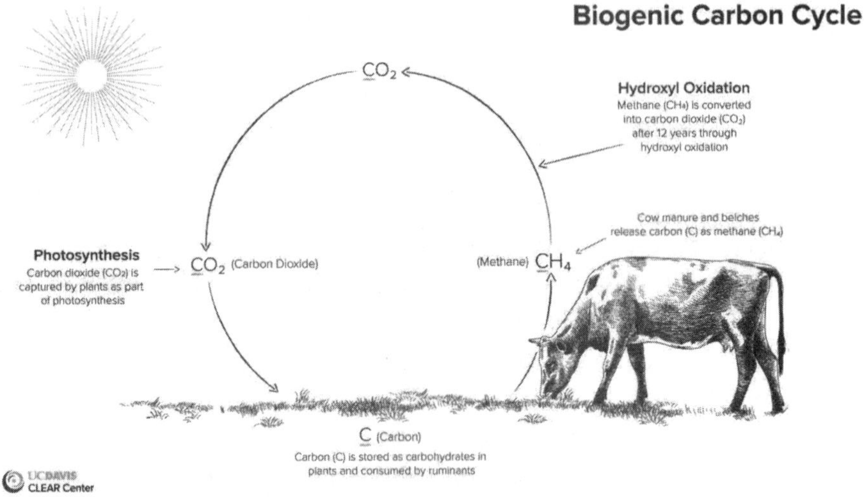

Biogenic Carbon Cycle

- **Photosynthesis**: Carbon dioxide (CO_2) is captured by plants as part of photosynthesis
- **Hydroxyl Oxidation**: Methane (CH_4) is converted into carbon dioxide (CO_2) after 12 years through hydroxyl oxidation
- Cow manure and belches release carbon (C) as methane (CH_4)
- CO_2 (Carbon Dioxide)
- (Methane) CH_4
- C (Carbon): Carbon (C) is stored as carbohydrates in plants and consumed by ruminants

© UC DAVIS CLEAR Center

(*Source: CLEAR Center, University of California, Davis*)

economy between plants and animals, as illustrated above: Plants absorb the carbon dioxide from the atmosphere, animals eat plants and fertilise the plants with their 'waste'. Herbivores have four stomachs—they eat and, through enteric fermentation, digest the carbon in the cellulose of the plant and transform it into milk and other metabolic processes.

Cars eat food

At the G20 Summit in New Delhi, on September 9, 2023, the Global Biofuel Alliance (GBA), comprising both biofuel producers and consumers, was launched by India. Member

---

water.unl.edu/article/manure-nutrient-management/reducing-carbon-footprint-cattle-operations-through-diet.

countries include India, Singapore, Bangladesh, Italy, the US, Brazil, Argentina, Mauritius and the UAE. Among the supporting organisations are the International Energy Agency (IEA), the International Civil Aviation Organization (ICAO), the WEF and World LPG Association.

According to predictions, the global ethanol market, which was US$99.06 billion in 2022, is expected to surpass US$162.12 billion by 2032.[33] The main sources of biofuels are sugarcane, grains and agricultural waste. Yet, industrial biofuels raise several questions: Are they truly carbon neutral? Will the poor gain or lose with an explosive increase in the production of industrial biofuels? What are the implications of industrial biofuels vis-à-vis soil, ecology, land sovereignty and food sovereignty?

Even though biofuels are being promoted as a source of clean renewable energy, there are ecological reasons why converting crops like soyabean, corn and palm into liquid fuels can actually aggravate the CO_2 burden, worsening the climate crisis while also contributing to the erosion of biodiversity and the depletion of water resources.

First, there is the question of deforestation in order to increase soyabean and palm oil cultivation. According to the FAO, 25 to 30 percent of the GHGs released in the atmosphere each year are a result of deforestation. Second, if we take into account the production of biofuels, we see that they cause more GHG emissions than conventional fuels. According to

---

[33] 'Ethanol Market Size to Hit around USD 162.12 Billion by 2032', Precedence Research, January 2024, https://www.precedenceresearch.com/ethanol-market.

two important studies published in February 2008,[34] ethanol and biodiesel production are linked to increased CO_2 emissions and destruction of biodiverse forests, as well as air and water pollution. The destruction of natural ecosystems, be it rainforests in the tropics or grasslands in South America, not only releases GHGs into the atmosphere, but also deprives the planet of natural sponges that absorb carbon emissions.

Ethanol production entails the use of 1,700 gallons of water to produce one gallon of ethanol. Corn, one of the sources, requires more nitrogen fertiliser, more insecticides and more herbicides than any other crop. Increased use of corn and soyabean oil for biofuel production has raised world food prices by 10 percent, according to an IMF report.[35] Biofuels have pushed up feedstock prices, too. Around the world, acreage under forests or food agriculture is being converted for biofuel production, thus aggravating the water situation. An International Water Management Institute study warned that ambitious plans in China and India to greatly increase domestic production of biofuels derived from crops will place enormous stress on these countries' water supply, seriously undermining

---

[34] Rhett A. Butler, 'Biofuels Are Worsening Global Warming', Mongabay, February 7, 2008, https://news.mongabay.com/2008/02/biofuels-are-worsening-global-warming. See also Joseph Fargione et al., 'Land Clearing and the Biofuel Carbon Debt', *Science* 319, no. 5867 (2008): 1235–8; Tim Searchinger et al., 'Use of U.S. Croplands for Biofuels Increases Greenhouse Gasses through Emissions from Land-Use Change', *Science* 319, no. 5867 (2008): 1238–40.

[35] Valerie Mercer-Blackman, Hossein Samiei, and Kevin Cheng, 'IMF Survey: Biofuel Demand Pushes Up Food Prices', IMF News, October 17, 2007, https://www.imf.org/en/News/Articles/2015/09/28/04/53/sores1017a.

The Fake Food Dystopia

their ability to meet food and feed demands.[36]

Energy can only be considered sustainable if it does not compete with the food supply, does not divert organic matter from maintaining the essential ecosystem and is decentralised.

Replacing one resource intensive and energy intensive industrial food system with another within the old paradigm is not the transformation of the food system that the people and the earth are seeking. Then, because we continue to lose control over the origin and production of food, we are gradually giving up our food sovereignty. Artificial food does not present itself as a clear alternative to our diet; rather, by disguising itself as a form of traditional food, it tries to sneak up to our tables. It is a full-fledged counterfeiting operation that aims to gain control over our diets by making food ever more dependent on the multinational companies that produce and patent it.

The earth and our food systems are becoming more vulnerable and more undemocratic as resources and power are extracted from both the earth community and food communities.

---

[36] 'Biofuel Threatens China, India Water Supply', International Water Management Institute, May 5, 2008, https://www.iwmi.cgiar.org/news/biofuel-threatens-china-india-water-supply.

6 | The Future of Food

The farm crisis

The year 2024 started with massive farmers' agitations across Europe, as well as in India. In Europe, the protests spread across the Netherlands, Poland, Germany, Lithuania, Romania, France, Scotland, Portugal, Spain, Italy, Greece, Britain, Bulgaria, Belgium, Ireland, the Czech Republic, Wales and Norway.

In India, just over two years after their thirteen-month-long protest forced a roll-back of controversial laws aimed at 'modernising' the agricultural sector—relaxing the rules around the sale, pricing and storage of farm produce that had protected farmers from an unfettered free market for decades—farmers from the states of Punjab, Haryana and Uttar Pradesh hit the streets to demand, among other things, a higher Minimum Support Price (MSP) for their crops, a debt waiver and withdrawal from WTO agreements.

Farmers everywhere are fighting for survival; they are resisting extinction. In 2001, six years after the WTO rules came into effect, I wrote a report entitled *Yoked to Death*.[1] The very industrial agriculture system that contributes to emissions and drives climate change has also defined farmers as a dispensable part of the food system. In the context of US agriculture, the American economist Kenneth Boulding observed, 'In this country we have moved from 90 percent of the population in

---

[1] Vandana Shiva, *Yoked to Death: Globalisation and Corporate Control of Agriculture* (New Delhi: Research Foundation for Science, Technology and Ecology, 2001).

The Future of Food

agriculture to 8 percent in 200 years.' (It is less than two percent now.) He added,

> The only way I know to get toothpaste out of a tube is to squeeze the tube, and the only way to get people out of agriculture is likewise to squeeze agriculture. It just has to be made less profitable than other occupations.[2]

Farmers are being squeezed out of agriculture on account of multiple pressures. Industrial agriculture forces farmers to rely on costly inputs. They are transformed from being producers of food to becoming consumers of expensive chemicals and industrial seeds with a false calculus of productivity, which I call pseudo productivity. Productivity is output per unit. But all outputs should be measured, not just the monoculture commodity that the market wants to extract.

Small farms are critical to reducing external energy use, which agronomist and entomologist David Pimentel, along with researcher and academic Mario Giampietro, call exosomatic energy. Differentiating between endosomatic and exosomatic energies, they explain,

> Endosomatic energy is generated through the metabolic transformation of food energy into muscle energy in the human body. Exosomatic energy is generated by transforming energy outside of the human body [by mechanical means], such as by burning oil in a tractor.

Pimentel and Giampietro found that it takes 10 kilocalories of exosomatic energy to produce one kilocalorie of food in the US.

---

[2] Kenneth Boulding, 'Agricultural Organizations and Policies: A Personal Evaluation', in Earl O. Heady et al., *Farm Goals in Conflict: Family Farm, Income, Freedom, Security* (Ames: Iowa State University Press, 1963), 157.

The remaining nine kilocalories go towards creating waste and pollution and increasing entropy. Part of this wasted energy is released into the atmosphere and contributes to climate change.[3]

Industrial agriculture in the US uses 380 times more energy per hectare to produce rice than a traditional farm in the Philippines; energy use per kilo of rice is 80 times greater in the US than in the Philippines. Corn production in the US requires 176 times more energy per hectare than on a traditional farm in Mexico, and 33 times more per kilo. One cow maintained and marketed in the industrial system requires 6 barrels of oil. And while 450 grams of breakfast cereal provides only 1,000 kilocalories of food energy, it uses 7,000 kilocalories of energy for processing.[4]

Physicist Amory Lovins uses the term 'energy slaves' to denote the hidden energy used in inefficient industrial systems and industrial societies. Energy slaves consume resources and energy, but cannot offer care to the earth. Way back in 1975, Lovins had stated that an average American had 250 times more energy slaves than a Nigerian.[5] With a population of over

---

[3] David Pimentel and Mario Giampietro, *Food, Land, Population and the U.S. Economy* (Washington, D.C.: Carrying Capacity Network, 1994). In 1994, Pimentel and Giampietro estimated the output/input ratio of agriculture to be around 1.4. For 0.7 kcal of fossil energy consumed, US agriculture produced 1 kcal of food; see 'How Much Oil Does US Agriculture Use?', Physics Forum, June 29, 2010, https://www.physicsforums.com/threads/agricultural-petroleum-use.413197.

[4] Agostinho Moniz and Gyaw Shine Oo, 'Analysis of Energy Use in Rice Production, Post-Production and Cooking, Laguna Philippines', *IOSR Journal of Agriculture and Veterinary Science* 16, no. 7 (July 2023): 4–12; David Pimentel, 'Energy Inputs in Food Crop Production in Developing and Developed Nations', *Energies* 2, no. 1 (2009): 1–24.

[5] Amory B. Lovins, *World Energy Strategies: Facts, Issues, and Options* (London: Friends of the Earth Ltd., 1975).

The Future of Food

7.7 billion people living under forced industrialisation and energy intensive digitalisation, the population of energy slaves today is more than 3.35 trillion.

Every step towards displacing real people and substituting them with 250 energy slaves is driving the climate crisis, the destruction of forests and biodiversity, the destitution of farmers and an unemployment crisis. Pseudo efficiency hides the full ecological footprint of a production system. It hides true costs through externalities and subsidies. It cherry-picks one tiny technology fragment from an entire system and presents it as more 'efficient', even though the system as a whole is crude, violent, inefficient and destructive.

What does all this mean in terms of feeding the world? It is clear that by substituting people with energy slaves, machines and chemicals, industrial agriculture is virtually eliminating small farms and small farmers' capacity to produce diverse outputs of nutritious crops which are critical to global food security. While using 75 percent of the land, industrial agriculture only produces 30 percent of the food we eat—small, biodiverse farms using 25 percent of the land provide the rest, 70 percent. If the share of industrial agriculture and industrial food in our diet were to increase to 45 percent, we would end up with a dead planet. No life. No food. Both from the point of view of food productivity and food entitlements, industrial agriculture is deficient. Protecting small farms which conserve biodiversity is thus a food security imperative.

How does the decline in output get translated into an increase? There are a number of strategies which allow this inversion to take place and an illusion of growth to be created. First, a monoculture paradigm looks only at one element of a system and treats an increase in one part as an increase in the whole system. For instance, by focussing only on yield increases of individual cereals like rice or wheat, the reduction in straw availability for fodder is externalised and not accounted for. The biodiversity of crops that disappear when diversity is displaced by monocultures is never part of the calculation. A second strategy is to exclude the higher inputs from the resource equation. Consequently, resource waste is not taken into account so that low resource-use productivity is converted into high-commodity productivity.

On the input side, a systems assessment should include fossil energy input, capital input, chemical input and water input. However, a pseudo productivity measure conceals the costly external inputs while only calculating labour input. In this pseudo productivity calculus, farmers have been reduced to being referred to as 'the denominator'. There is rich talk about reducing the number of farmers in order to increase productivity and growth—but farmers are not an input; they are members of society. A society that values the health of the planet and of its people ensures that the producers of food, our most basic need, the basis of our health and the basis of our freedom are treated with respect, fairness and justice. Farmers are at the heart of the food sovereignty of communities and countries.

The corporate hijack of the food system

In the eight years following the establishment of the WTO, Indian farmers lost US$26 billion annually due to falling prices.[6]

---

[6] Vandana Shiva, Afsar H. Jafri, and Kunwar Jalees, *The Mirage of Market Access: How Globalisation is Destroying Farmers' Lives and Livelihoods* (New Delhi: Navdanya/RFSTE, 2003).

The Future of Food

Farming incomes have been steadily collapsing everywhere, which is why, time and again, farmers have risen against WTO rules and are demanding that food and agriculture be withdrawn from the WTO. In India, the issue of fair and liveable incomes has been articulated through demanding a legally binding MSP for produce. In Europe, too, farmers' protests centre on justice and fair incomes in the context of globalisation and corporate control over food and farming. According to Welsh farmer Ioan Humphreys, 'We as farmers are the heart of this country, but at this moment in time we're treated like the appendix.' Yorkshire farmer Anna Longthorp says,

> The unfairness in the supply chain has been allowed to continue for far too long, with all the risk and cost pressures dumped on farmers and no control over the pricey recipe. All the time corporate middlemen and supermarkets make huge returns whilst farmers subsidise profits. It is only a matter of time before the top heavy food system buckles and farmers decide enough is enough—that's where we are now.[7]

The extremely high ecological, economic and social externalities of industrial farming along with the globalised corporate food system have been borne by the earth through climate change and species extinction. They are also being borne by farmers worldwide as they, too, slowly become threatened by extinction. Declining farm incomes have already crossed the survival threshold. This is the subsidy corporations are extracting from farmers, who are paying with their very lives.

The globalised industrial food system is an extractive economy that leaches the last bit of fertility from the soil, the

---

[7] No Farmers, No Food is a protest campaign launched by farmers in the UK. For more, see https://twitter.com/NoFarmsNoFoods.

THE NATURE OF NATURE

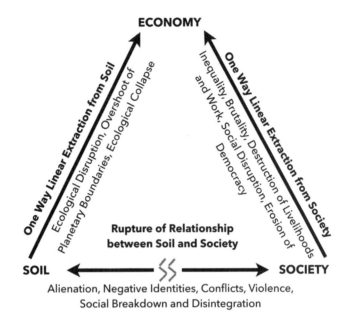

The globalised industrial food system is indifferent to the instabilities and vulnerabilities it creates for the planet and people.
(Source: Navdanya International)

last drop of water from the earth, the last seed from farmers, and self-organisation and health from the planet and people. Concentration and centralisation, distancing and separation of food producers from consumers and data from knowledge, are creating an inverted pyramid of unstable power which can topple as a result of small perturbations.[8]

---

[8] Vandana Shiva and Kartikey Shiva, with Neha Raj Singh, *The Future of Our Daily Bread: Regeneration or Collapse?* (New Delhi: Navdanya International/ RFSTE, 2018), https://seedfreedom.info/wp-content/uploads/2018/11/The-Future-of-Our-Daily-Bread-_-LowRes-_-19-11-2018-REVISED.pdf.

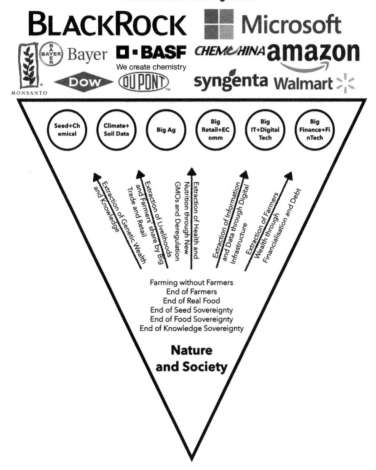

An extractive economy erodes the ecological and social foundations of food in nature and society.
(Source: Navdanya, https://www.navdanya.org/living-food/colonialism,-globalisation-are-at-the-roots-of-ecological-emergency,-farmers-distress)

Digital agriculture

Industrial agriculture envisions the future of farming with spyware and with surveillance drones collecting data, removing food even further from the ecological web of life.

Digital agriculture, based on Big Data and supported by the Agriculture Innovation Mission for Climate, or AIM4C, is being promoted aggressively to spread the false notion that farming without farmers, without nature's and farmers' intelligence, is the only climate-positive way forward. Launched in 2021, this joint initiative by the UAE and USA focuses on boosting 'data-driven technologies' and 'science-based decision-making' in farming. So far, 45 countries, including India, have come on board to support 'precision agriculture' and 'dramatically [ramp up] investment in and support for climate-smart agriculture and food systems'.[9]

But digital agriculture is simply the next phase of centralised, corporate-controlled industrial agriculture. It will aggravate the climate crisis, not solve it, because it entrenches the use of external inputs created by the very corporations that have contributed to climate change in the first place. For instance, spraying harmful pesticides from high-tech drones will not reduce the pesticide burden; rather, it will be spread over a wider area, even to farms that want to remain pesticide-free. Digitalisation will also increase energy use for computational processes.[10]

---

[9] 'AIM for Climate Summit 2023 Opens With New Agriculture Initiatives', Bio.News, May 9, 2023, https://bio.news/climate-change/aim-for-climate-summit-2023-opens-with-new-agriculture-initiatives.

[10] Steven Gonzalez Monserrate, 'The Staggering Ecological Impacts of Computation and the Cloud', MIT Press Reader, February 14, 2022, https://thereader.mitpress.mit.edu/the-staggering-ecological-impacts-of-computation-and-the-cloud.

The Future of Food

We are being told that Big Data, not farmers, will feed us. The reality is that Big Data is nothing but data mined from farmers, processed through algorithms and then sold back to them as additional external input. In 2013, Monsanto acquired the world's largest climate data corporation, the Climate Corporation, for US$1 billion. In 2014, it bought the world's largest soil data corporation, Solum Inc., 'to provide farmers insights and data of their fields based on historical crop, field, and weather data'.[11]

Earlier, the Poison Cartel forced farmers to buy external chemical inputs like GMO seeds and synthetic fertilisers; now they are forcing them to buy knowledge. Data is the new oil, the new fossil fuel, that is making agriculture non-renewable by reducing food and knowledge to scarce resources that must be procured at a high cost. But data is not knowledge; it is merely a commodity that makes farmers less connected with the earth by outsourcing their traditional knowledge base to Big Agribusiness. Liam Condon, member of the board of management of Bayer AG and president of the company's Crop Science Division, made the following claim in a piece in *POLITICO*:

> We know that innovation is the key to farming's future, and we are already seeing how new technologies—including digital farming, precision agriculture and plant breeding tools—are lessening agriculture's ecological footprint.... And today more farmers are using the latest in digital analytics to help them prevent problems before they start, so that they can grow more food while using less land, energy and water.[12]

---

[11] Wikipedia, s.v. 'The Climate Corporation', last modified May 5, 2023, https://en.wikipedia.org/wiki/The_Climate_Corporation. It's been a decade since Monsanto (now Bayer) bought out the Climate Corporation and Solum Inc. There is no further information on the impact of these acquisitions.

[12] Liam Condon, 'Farming's Future Belongs to All of Us', POLITICO, September 26, 2019, https://www.politico.eu/sponsored-content/farmings-future-belongs-to-all-of-us.

In a bid to expand digital agriculture in farming, a new legislation has been introduced in the US: the Linking Access to Spur Technology for Agriculture Connectivity in Rural Environments Act of 2023, or LAST ACRE Act, will create a US Department of Agriculture (USDA) Rural Development Grant programme to deliver broadband Internet to farms 'to advance precision agriculture connectivity nationwide'.

The Poison Cartel, which became the biotech industry, is converging with Big Tech like Microsoft and Facebook as well as Big Fintech, including asset management companies like BlackRock and Vanguard, to forge a digital, genetic and chemical dictatorship over food. Bill Gates is working closely with Bayer and the Poison Cartel to create one centralised, industrial, toxic, energy intensive digital agriculture monoculture for the entire world through his dystopic vision of Gates Ag One,[13] whose purpose is not to provide fresh food to local communities but to produce raw material for lab food. I call this food totalitarianism.

The corporate control of our food systems is leading to new levels of concentration of economic power, both through mergers within each sector and through technological integration across sectors, with biotechnology, information and digital and financial technologies becoming a single technological continuum. It is further distancing and separating food from its sources in seed, soil and water, and in the creative contributions of small farmers who invest in care for the earth.

---

[13] Navdanya, 'Gates Ag One: The Recolonisation of Agriculture', Independent Science News for Food and Agriculture, November 16, 2020, https://www.independentsciencenews.org/commentaries/gates-ag-one-the-recolonisation-of-agriculture.

It is distancing 'data' from reality and displacing real knowledge rooted in experience, practice, care and diverse intelligences which allows for choices that guide evolution and contribute to resilience. Separated from ecological and social systems, food becomes non-food, anti-food and fake food.

Diversity and decentred self-organising are the characteristics of living systems and their resilience, including living seed, living soil, living food, living knowledges, living economies and living democracies. External inputs, external control, uniformity, monocultures, centralisation and concentration create vulnerability and unstable, non-sustainable systems prone to break down. We need to grow the economy of care and shrink the economy of greed.

The future of food

Rejuvenating and regenerating the planet through ecological processes is a survival imperative for the human species and all living beings. Working with the earth and her systems and processes, with her plants and biodiversity, we create negative entropy and reduce pollution and waste to zero.

Agriculture is the science and art of cultivating the earth, including the harvesting of crops and the rearing and management of livestock. It is the culture of the land, in accordance with the ecological principles of nature: 1) self-organisation and autopoiesis; 2) diversity, not monocultures and uniformity; 3) symbiosis; 4) the law of return; and 5) sharing the earth's gifts in the commons. These principles have created food systems that have lasted for centuries because they walk the path of life as laid down by nature, respecting nature's circular economies on which all life depends. Practiced by diverse schools of ecological agriculture—organic farming, permaculture, biodynamic farming, natural farming—these principles are referred to as agroecology.

The two fundamental ecological cycles are nutrient and water. The living carbon cycle is a food cycle, a nutrient cycle. Maximising biodiversity and biomass density not only produces more nutrition per acre through increased photosynthesis, thereby addressing food insecurity, but also increases the living carbon in the soil; boosts nutrients, including nitrogen, magnesium, zinc and iron; and enhances the density of beneficial organisms. This is farming with integrity.

Biodiversity regeneration and intensification allow us to grow diverse, healthy food. The more density we introduce in the ecological circular nutrient economy in partnership with nature's recycling, the more fertile our soils will become. Greater organic matter available to return to the soil will reverse desertification, which is among the primary reasons for the displacement and uprooting of people. Additionally, the more plants we have, the more they will recycle and fix atmospheric carbon and nitrogen, reducing both harmful emissions and the stock of pollutants in the air.

The symbiotic flow of nutrients between the biosphere and atmosphere heals broken climate cycles; the flow of biodiversity and its nourishment from the soil to our gut microbiome heals our broken health.[14] Thus, regenerating the biodiversity of plants, and of the soil, is the real solution to climate change.

---

[14] Fiona Harvey, 'Improving Soil Could Keep World Within 1.5C Heating Target, Research Suggests', *The Guardian*, July 4, 2023, https://www.theguardian.com/environment/2023/jul/04/improving-farming-soil

The Future of Food

Climate action is de-addiction from fossil fuels, fossil chemicals and an industrial infrastructure. The first step towards this is to change our way of thinking from a dead earth paradigm to a living planet paradigm. In the process, we derive the power to act, to co-create and co-produce with the earth; we move from paradigms and economic systems that create scarcity and sickness to paradigms and economic systems that create abundance and well-being for all humans and all species.

Living systems have negative feedback loops which keep the conditions of the planet within boundaries that are favourable to life. Climate change is a result of the rupturing of these boundaries. Transitioning to organic agriculture has the potential to remove 100 percent GHGs from the atmosphere while also regenerating the soil-food web which is the source of recycling nutrients, including the nutrient cycles that connect soils and plants to the atmosphere.

What governments and policymakers need to do is to stop promoting the oil-based paradigm and advocate, instead, an earth-centred, soil-based way of thinking; they need to end subsidies to fossil fuels and industrial agriculture and support communities in a transition to local, ecological, biodiverse, poison-free and fossil fuel-free food and farming systems.

Agriculture can shift from being a major emitter of GHGs to becoming the most significant mitigator of emissions by following nature's processes of biodiversity and photosynthesis intensification. André Leu of Regeneration International

---

-carbon-store-global-heating-target; André Leu, 'Maximizing Photosynthesis and Root Exudates through Regenerative Agriculture to Increase Soil Organic Carbon to Mitigate Climate Change', *SCIREA Journal of Agriculture* 8, no. 1 (February 2023), https://article.scirea.org/pdf/210317.pdf.

explains, on the basis of scientific evidence, that the system of photosynthesis maximisation and of returning carbon to the soil

> could sequester 34.74 Gt [gigatonnes] of CO_2 per year. This is more than the current anthropogenic emissions of 26.88 Gt of CO_2 eq per year and would achieve negative emissions. . . . Scaling up 10 % of various. . . regenerative agriculture systems. . . can significantly contribute to achieving the negative emissions needed to limit global warming to 1.5°C higher than pre-industrial levels.[15]

Marine biologist and environmental informatics professor Dr Jacqueline McGlade affirms that

> using better farming techniques to store 1% more carbon in about half of the world's agricultural soils would be enough to absorb about 31 gigatonnes of carbon dioxide a year. . .[which is] not far off the 32 gigatonnes gap between current planned emissions reduction globally per year and the amount of carbon that must be cut by 2030 to stay within 1.5C.[16]

Organic soils are rich in mycorrhizal fungi which are nutrients for plants, while plants deliver carbohydrates as food. The symbiotic relationship between plants and fungi is the basis of our food system. Global plant communities contribute 13.12 Gt of CO_2 to diverse species of mycorrhizal fungi, which is approximately 36 percent of current CO_2 emissions from fossil fuels.[17]

The filaments of the fungi are microscopic, yet they hold 36 percent of the world's annual carbon emissions from fossil

---

[15] Leu, op. cit.

[16] Findings of research by McGlade, quoted in Harvey, op. cit.

[17] Brian Owens, 'Underground Fungi Absorb Up to a Third of Our Fossil Fuel Emissions', *New Scientist*, June 5, 2023, https://www.newscientist.com/article/2376827-underground-fungi-absorb-up-to-a-third-of-our-fossil-fuel

fuels.[18] Research conducted by Navdanya shows that fungi in organic soils increase 36-fold, while in chemical soils they decline by up to 49.7 percent.[19] Bacteria increase sixfold. So, organically-farmed soils host substantially greater microbial biomass, microbial activity and diversity than conventionally-farmed soils. Specifically, the microbial biomass of carbon and nitrogen is 41 percent and 51 percent higher, respectively, on organic plots, which also has 32 percent to 74 percent more microbial enzyme activity. Most studies reviewed reveal significant differences in the composition of the soil community, with organic farming practices increasing the abundance and activity of soil life.[20]

-emissions/; Heidi-Jayne Hawkins et al., 'Mycorrhizal Mycelium as a Global Carbon Pool', *Current Biology* 33, no. 11 (June 2023): R560–R573, https://doi.org/10.1016/j.cub.2023.02.027; Jeff Powell et al., 'Hidden Carbon: Fungi and Their "Necromass" Absorb One-Third of the Carbon Emitted by Burning Fossil Fuels Every Year', The Conversation, Yahoo! News, June 5, 2023, https://au.news.yahoo.com/hidden-carbon-fungi-necromass-absorb-200327075.html.

[18] Liz Kimbrough, 'Mycorrhizal Fungi Hold CO_2 Equivalent to a Third of Global Fossil Fuel Emissions', Mongabay, June 13, 2023, https://news.mongabay.com/2023/06/mycorrhizal-fungi-hold-co2-equivalent-to-a-third-of-global-fossil-fuel-emissions.

[19] I. Rathore, V. Shiva, E. Thomas et al., 'A Comparison on Soil Biological Health on Continuous Organic and Inorganic Farming', *Horticulture International Journal* 2, no. 5 (2018): 256–62, https://medcraveonline.com/HIJ/a-comparison-on-soil-biological-health-on-continuous-organic-and-inorganic-farming.html.

[20] David R. Montgomery and Anne Biklé, 'Soil Health and Nutrient Density: Beyond Organic vs. Conventional Farming', *Frontiers in Sustainable Food Systems* 5 (November 2021), https://www.frontiersin.org/articles/10.3389/fsufs.2021.699147/full#B74; M. Lori et al., 'Organic Farming Enhances Soil Microbial Abundance and Activity—A Meta-Analysis and Meta-Regression', *PLoS ONE* 12, no. 7 (2017): e0180442, https://www.doi.org/10.1371/journal.pone.0180442.

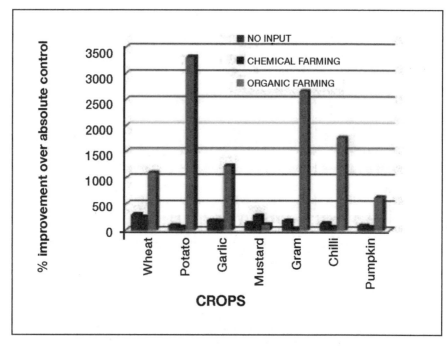

Status of fungal population under different crops and input in organic and chemical farms.

(Source: https://medcraveonline.com/HIJ/a-comparison-on-soil-biological-health-on-continuous-organic-and-inorganic-farming.html)

Biodiversity provides multiple ecological functions that enable farming without life-destroying synthetic chemicals and fossil fuels. It can reduce emissions by 20 percent while also protecting forests, the earth's lungs. At the same time, shifting from globalisation to localisation of food systems can reduce emissions from 'food miles'; shifting from industrial to artisanal processing, and from ultra-processed to fresh foods, can also reduce emissions. Consumption of fresh food can reduce plastic

Carbon is captured by plants through photosynthesis. Some of this carbon then goes into the networks of mycorrhizal fungi. These fungi also release some of it as CO_2 and as compounds into the soil.

(Source: Adam Frew/Author provided using BioRender)

and aluminium packaging, and together, these changes can reduce another 20 percent emissions.[21]

---

[21] 'Food and Climate Change: The Forgotten Link', GRAIN, September 28, 2011, https://grain.org/en/article/4357-food-and-climate-change-the-forgotten-link. These inferences are based on removing the contributions

Apart from lowering emissions, ecological food systems also contribute to negative emissions. They help draw down the excess CO_2 through photosynthesis and locate carbon where it belongs—as living carbon, the molecule of life, in plants and as humus in soil. We need to remind ourselves that living carbon gives life and that dead fossil carbon disrupts living processes.

Dead carbon must be left underground—this is an ethical obligation and an ecological imperative. This is why 'decarbonisation' without qualification and a distinction between living and dead carbon is scientifically and ecologically inappropriate. If we decarbonised the economy, we would have no plants and therefore no life on earth which creates, and is sustained by, living carbon. A decarbonised planet is a dead planet. What we need to do is to *recarbonise* the world with living carbon, and *decarbonise* it of dead carbon; we need to see new potential in an economy of care for the soil.

Healthy soils, healthy people

The transition for climate action is a transition from oil-based thinking and living to soil-based thinking and living. Healthy soils are dense with biodiversity. One gram of organic soil contains 30,000 protozoa, 50,000 algae, 400,000 fungi. One teaspoon of living soil contains one billion bacteria, which translates to one tonne per acre. One square cubic metre of soil contains 1,000 earthworms, 50,000 insects, 12 trillion roundworms. Living soils are the biggest reservoir of both water and nourishment; soils rich in humus can hold 90 percent of their weight in water.

---

of the fossil fuel and fossil chemical farming system mentioned in the GRAIN article, with about more than 50 percent emissions emanating from the fossil food system.

Nutrient	Change under Chemical Farming	Change under Organic Farming
Organic Matter	-14%	+29-99%
Total Nitrogen (N2)	-7-22%	+21-100%
Available Phosphorous (P)	0%	+63%
Available Potassium (K)	-22%	+14-84%
Zinc (Z)	-15.9-37.8%	+1.3-14.3%
Copper (Cu)	-4.2-21.3%	+9.4%
Manganese (Mn)	-4.2-17.6%	+14.5%
Iron (Fe)	-4.3-12%	+1%

Nutrition comparison between chemical farming vs. organic farming in continuously farmed soils in Navdanya farms in Uttarakhand over twenty years. The comparative study was published in 2018.[22]

(Source: Navdanya)

On the Navdanya farm in Dehradun, in the foothills of the Himalayas, organic matter has increased by up to 99 percent; zinc has increased 14 percent; and magnesium by 14 percent. We did not add these externally; they have been produced by the billions of soil microorganisms that are found in living soils.

When we add urea to soil, this rich biodiversity of soil microorganisms that creates the diversity of soil nutrients is destroyed. Similarly, when we eat poisons, our gut microbiome begins to get 'desertified'. With 'humus' and 'human' having the

---

[22] Rathore et al., op. cit.

same etymology, is it any wonder that when soils are sick and desertified, societies become sick, too? Since we are more bacteria than human, when the poisons used in industrial agriculture (such as pesticides and herbicides) reach our gut through food, they can kill beneficial bacteria essential to good health.

There is an intimate connection between the soil, plants, our gut and our brain. Emeran Mayer observes in *The Mind-Gut Connection*:

> For decades, the mechanistic, militaristic disease model set the agenda for medical research: As long as you could fix the affected machine part, we thought, the problem would be solved; there was no need to understand its ultimate cause. . . . We are just beginning to realise that the gut, the microbes living in it—the gut microbiota—and the signaling molecules that they produce from their vast number of genes—the microbiome—constitute one of the major components of these regulatory systems.[23]

The gut has its own nervous system, often referred to as the enteric nervous system, or ENS, with 50–100 million nerve cells. The trillions of bacteria in the gut emanate their own intelligence. American biologist James Shapiro says,

> Bacteria possess many cognitive, computational and evolutionary capabilities. . . . [Studies] show that bacteria utilize sophisticated mechanisms for intercellular communication and even have the ability to commandeer the basic cell biology of 'higher' plants and animals to meet their own needs. This remarkable series of observations requires us to revise basic

---

[23] Emeran Mayer, *The Mind-Gut Connection: How the Hidden Conversation Within Our Bodies Impacts Our Mood, Our Choices, and Our Overall Health* (New York: Harper Wave, 2016), 6, 4.

ideas about biological information processing and recognize that even the smallest cells are sentient beings.[24]

It is because bacteria are sentient that they develop resistance to antibiotics. Monsanto says its herbicide, Roundup, which is used as a weed killer, is safe for humans because humans do not have the shikimate pathway through which some plants and protozoa, bacteria, fungi and algae biosynthesise folates and aromatic amino acids. But the bacteria in our gut do have the shikimate pathway, and they are being killed by herbicides like Roundup, leading to serious disease epidemics, from increasing intestinal disorders to neurological problems. The bacteria in our gut produce three aromatic amino acids: tryptophan, tyrosine and phenylalanine, all through the shikimate pathway. Since our cells do not have this pathway, they themselves are unable to make these nutrients, so we depend on gut bacteria. These essential amino acids are precursors to the neurotransmitters dopamine, serotonin, melatonin and adrenaline, as well as thyroid hormone, folate and vitamin E. The destruction of gut bacteria can lead to deficiencies in these important biological molecules, impairing neurological functions. Specific molecules and phytochemicals found in herbs and spices activate specific taste receptors and trigger particular metabolic processes. Mayer writes,

> The gut's elaborate sensory systems are the National Security Agency of the human body, gathering information from all areas of the digestive system, including the esophagus,

---

[24] James A. Shapiro, 'Bacteria are Small but Not Stupid: Cognition, Natural Genetic Engineering, and Socio-bacteriology', *Stud in Hist Philos Biol Biomed Sci* 38, no. 4 (December 2007): 807–19, https://www.doi.org/10.1016/j.shpsc.2007.09.010.

stomach, and intestine, ignoring the great majority of signals, but triggering alarm when something looks suspicious or goes wrong. As it turns out, it's one of the most complex sensory organs of the body.[25]

Eating, then, is a conversation between the soil, plants, the cells in our gut and the cells in our food, and between gut and brain. Eating is an intelligent act at the deepest cellular and microbial levels. Cellular communication is the basis of health and well-being, and also the root of disease. We may be ignorant about the links between food and health, but our cells know them well. Since food carries in it the memory of the biodiversity in the soil and plants, how food is grown is a major determinant of health.

We are 90 percent other beings, primarily our fellow microbes that keep us healthy. The human microbiome consists of all the microbes—bacteria, fungi and viruses—that live within us or on us, including the skin, mammary glands, seminal fluid, uterus, ovarian follicles, lung, saliva, oral mucosa, conjunctiva, biliary tract and gastrointestinal tract. It has been estimated that there are over 380 trillion viruses that inhabit us, a community collectively known as the human virome. Our gut is a microbiome which contains trillions of bacteria. There are 100,000 times more microbes in our gut than people on the planet. To function in a healthy way, the gut microbiome needs a diverse diet, and a diverse diet needs diversity in our fields and gardens.

The notion of 'one health' calls for an integrated, ecological approach, not a militarised, mechanistic, reductionist approach of declaring species that are part of our body and

---

[25] Mayer, op. cit., 63.

part of the earth as enemies to be exterminated with biocides, insecticides, herbicides and fungicides. One health recognises our entanglement with other beings. We are interbeings, multi-species organisms, an ecological community, members of a complex, self-organised, self-regulated ecosystem.

We are facing an existential crisis with multiple emergencies—the health pandemic; the hunger pandemic; the poverty pandemic; the pandemic of fear and hopelessness, of inequality and disposability; the climate emergency; the biodiversity and extinction emergency. These emergencies are interconnected and have common roots in a mechanistic ontology and in an epistemology of separation.

The time for climate change denial and its disastrous consequences is long over. What we eat, how we grow the food we eat, how we distribute it, will determine whether humanity survives or pushes itself and other species to extinction. In the end, artificial, synthetic foods dismantle our connection with nature, and in so doing, they completely disregard the role of natural processes and the laws of ecology that are at the heart of real food production. Contrary to the claims of the agro-industry and food tech companies, food cannot be reduced to a commodity to be put together mechanically and artificially in labs and factories. Food is the currency of life, and it holds the contribution of all beings involved at all stages of production.

The world is witnessing a resurgence and reaffirmation of chemical-free, biodiverse and ecological agriculture, as practiced by small farmers. More and more people are adopting agroecology either in public lots or in private gardens. Jim Thomas, in his article 'George and the Food System Dragon', quotes the US Department of Agriculture ('not known for its romantic food sovereignty bias') as stating,

Around 15% of the world's food is now grown in urban areas. City and suburban agriculture take the form of backyard, rooftop and balcony gardening, community gardening in vacant lots and parks, roadside urban fringe agriculture and livestock grazing in open space. [26]

This is how food democracy and food sovereignty are being re-established, reclaimed from corporate control; this food system, free of poisons and plastics and based on locality, is nurturing both for our planet and for its denizens.

Animals, humans and nature have always lived in interconnected, symbiotic relationships which, in turn, regenerate all systems that support life. This synergy is vital to the renewal of soil fertility, the creation of habitat for biodiversity and the rejuvenation of the earth's water, carbon and nutrient cycles. The real solution does not lie in creating substitutes for food; it lies in understanding the needs of the ecosystems we are embedded in.

Earth care is action for climate justice, action for food justice, action for health justice and for social and economic justice for producers as well as consumers. With every seed we sow, every plant we grow, every morsel we eat, we make a choice between degeneration and regeneration.

---

[26] Jim Thomas, 'George and the Food System Dragon', Scan The Horizon, October 26, 2023, https://www.scanthehorizon.org/p/george-and-the-foodsystem-dragon.

Acknowledgements

Gratitude to my colleague Maya Goburdhun, who helped edit a very early draft of this book. To my colleagues in Navdanya International, for research and illustrations on fake food. To Ritu Menon, my publisher of four decades.

About the Author

Vandana Shiva is a world-renowned ecologist and a leader in the International Forum on Globalization and the Slow Food Movement. Director of Navdanya and of the Research Foundation for Science, Technology and Ecology and a tireless crusader for farmers', peasants' and women's rights, she is the author and editor of a score of influential books, among them *Terra Viva*; *Oneness vs. the 1%*; *Making Peace with the Earth*; *Soil Not Oil*; *Globalization's New Wars*; *Seed Sovereignty, Food Security*; and *Who Really Feeds the World?*

Shiva is the recipient of over twenty international awards, including the Right Livelihood Award (1993); the Medal of the Presidency of the Italian Republic (1998); the Horizon 3000 Award (Austria, 2001); the John Lennon-Yoko Ono Grant for Peace (2008); the Save the World Award (2009); the Sydney Peace Prize (2010); the Calgary Peace Prize (2011); and the Thomas Merton Award (2011). She was the Fukuoka Grand Prize Laureate in 2012.